T0328274

ENHANCING DISASTER
PREPAREDNESS

———

ENHANCING DISASTER PREPAREDNESS

FROM HUMANITARIAN ARCHITECTURE TO COMMUNITY RESILIENCE

Edited by

A. Nuno Martins
CIAUD, Research Centre for Architecture, Urbanism and Design,
Faculty of Architecture, University of Lisbon, Lisbon, Portugal

Mahmood Fayazi
The Institute for Disaster Management and Reconstruction,
Sichuan University and the Hong Kong Polytechnic University, Chengdu, Sichuan, China

Faten Kikano
Faculté de l'Aménagement, Université de Montréal, Montreal, Quebec, Canada

Liliane Hobeica
RISKam (Research group on Environmental Hazard and Risk Assessment and Management),
Centre for Geographical Studies, University of Lisbon, Lisbon, Portugal

ELSEVIER

Elsevier
Radarweg 29, PO Box 211, 1000 AE Amsterdam, Netherlands
The Boulevard, Langford Lane, Kidlington, Oxford OX5 1GB, United Kingdom
50 Hampshire Street, 5th Floor, Cambridge, MA 02139, United States

Notices
Knowledge and best practice in this field are constantly changing. As new research and experience broaden
our understanding, changes in research methods, professional practices, or medical treatment may become
necessary.

Practitioners and researchers must always rely on their own experience and knowledge in evaluating and
using any information, methods, compounds, or experiments described herein. In using such information or
methods they should be mindful of their own safety and the safety of others, including parties for whom
they have a professional responsibility.

To the fullest extent of the law, neither the Publisher nor the authors, contributors, or editors, assume any
liability for any injury and/or damage to persons or property as a matter of products liability, negligence or
otherwise, or from any use or operation of any methods, products, instructions, or ideas contained in the
material herein.

Library of Congress Cataloging-in-Publication Data
A catalog record for this book is available from the Library of Congress

British Library Cataloguing-in-Publication Data
A catalogue record for this book is available from the British Library

ISBN: 978-0-12-819078-4

For information on all Elsevier publications visit our website at
https://www.elsevier.com/books-and-journals

Publisher: Candice Janco
Acquisitions Editor: Marisa LaFleur
Editorial Project Manager: Pat Gonzalez
Production Project Manager: Bharatwaj Varatharajan
Cover Designer: Matthew Limbert

Typeset by TNQ Technologies

Working together
to grow libraries in
developing countries

www.elsevier.com • www.bookaid.org

Contents

Contributors

Kristjana Adalgeirsdottir
Aalto University, Helsinki, Finland

Lara Alshawawreh
Faculty of Engineering, Mutah University, Karak, Jordan

Anouck Andriessen
KU Leuven, Leuven, Belgium

Aditya Barve
Massachusetts Institute of Technology, Urban Risk Lab, Cambridge, MA, United States

Pablo Benetti
Post-Graduation Program in Urbanism, Faculty of Architecture and Urbanism, Federal University of Rio de Janeiro, Rio de Janeiro, Brazil

Lisa Bornstein
School of Urban Planning, McGill University, Montreal, Quebec, Canada

Liz Brogden
Queensland University of Technology, Brisbane, Queensland, Australia

Georgia Cardosi
Faculté de l'Aménagement, Université de Montréal, Montreal, Quebec, Canada

Sandra Carrasco
Faculty of Architecture, Building, and Planning, University of Melbourne, Melbourne, Victoria, Australia

Solange Carvalho
Post-Graduation Program in Urbanism, Faculty of Architecture and Urbanism, Federal University of Rio de Janeiro, Rio de Janeiro, Brazil

Raquel Colacios
School of Architecture, Universitat Internacional de Catalunya, Barcelona, Spain

Mauro Cossu
Faculté de l'Aménagement, Université de Montréal, Montreal, Quebec, Canada

Diane E. Davis
Harvard University Graduate School of Design, Cambridge, MA, United States

Mahmood Fayazi
The Institute for Disaster Management and Reconstruction, Sichuan University and the Hong Kong Polytechnic University, Chengdu, Sichuan, China

Eefje Hendriks
Department of Architecture, Eindhoven University of Technology, Eindhoven, The Netherlands

Adib Hobeica
Independent consultant, Coimbra, Portugal

Liliane Hobeica
RISKam (Research group on Environmental Hazard and Risk Assessment and Management), Centre for Geographical Studies, University of Lisbon, Lisbon, Portugal

Rosemary Kennedy
SubTropical Cities Consultancy, Brisbane, Queensland, Australia

Susan N. Kibue
Department of Architecture, Jomo Kenyatta University of Agriculture and Technology, Juja, Nairobi, Kenya

Faten Kikano
Faculté de l'Aménagement, Université de Montréal, Montreal, Quebec, Canada

Chetan Krishna
Massachusetts Institute of Technology, Urban Risk Lab, Cambridge, MA, United States

Jia Lu
University of Toronto, Centre for Landscape Research, Toronto, Ontario, Canada

Kelly Leilani Main
Massachusetts Institute of Technology, Urban Risk Lab, Cambridge, MA, United States

A. Nuno Martins
CIAUD, Research Centre for Architecture, Urbanism and Design, Faculty of Architecture, University of Lisbon, Lisbon, Portugal

Fadi Masoud
University of Toronto, Centre for Landscape Research, Toronto, Ontario, Canada

Miho Mazereeuw
Massachusetts Institute of Technology, Urban Risk Lab, Cambridge, MA, United States

David O'Brien
Faculty of Architecture, Building, and Planning, University of Melbourne, Melbourne, Victoria, Australia

Mayank Ojha
Massachusetts Institute of Technology, Urban Risk Lab, Cambridge, MA, United States

Angeliki Paidakaki
KU Leuven, Leuven, Belgium

Judith I. Rodríguez Portieles
Department of Architecture, Harvard University Graduate School of Design, Cambridge, MA, United States

Helena Sandman
Aalto University, Helsinki, Finland

Benjamin Schep
BuildtoImpact, Group 5 Consulting Engineers B.V., The Hague, The Netherlands

Miia Suomela
Aalto University, Helsinki, Finland

Cynthia Susilo
KU Leuven, Leuven, Belgium

Pieter van den Broeck
KU Leuven, Leuven, Belgium

Alexander van Leersum
BuildtoImpact, Group 5 Consulting Engineers B.V., The Hague, The Netherlands

Foreword

Diane E. Davis

Harvard University Graduate School of Design, Cambridge, MA, United States

The multiplying disasters of the 21st century remind us that we live in a world at risk. Vulnerabilities associated with environmental degradation, climate change and extreme weather events, earthquakes, health epidemics, and civil wars or other violent conflicts have destabilized fragile ecosystems, intensified social inequalities, fueled migration, and thrown the most disadvantaged among us into a downward spiral of precarity. The global COVID-19 pandemic is merely the latest disaster to drive this point home. The "risk society" that Ulrich Beck warned about in the 1990s is our reality. Recurring disaster is now normalized. To the extent that no part of the world can escape this future, risk may end up being the great global equalizer—even if social classes, neighborhoods, cities, and nations will continue to unevenly experience the effects and trauma of any given disaster. Although the traumatic impacts of a single disaster event may be hard to predict in form, timing, and intensity, the stark reality is that there will always be another. This is exactly why disaster preparedness must be the new modus operandi of our times, and why this volume arrives at the right moment.

The contributions in these pages can be seen as part of an essential toolkit for the risk society, as originally conceptualized by Beck (1992, p. 21), because they do offer a "way of dealing with hazards and insecurities induced and introduced by modernisation itself." Focusing on naturally triggered and human-made disaster contexts, and with an understanding of the spatial and social complexities that mediate any given disaster's overall impacts, the authors in this collective volume review both successful and failed strategies to recover from disastrous events, as well as initiatives related to disaster risk reduction. Drawing on case studies from around the world, they examine the role of design thinking, community involvement, government policies, and expert interventions in enabling or constraining effective disaster preparedness and response. The larger aim of any close analysis of what works and what does not is to be able to learn from mistakes, so as to be prepared to confront the next disaster down the road. This volume takes a much-needed step in that direction.

Yet, because the array of disasters examined here is relatively broad—ranging from housing issues in the context of forced displacement, hurricanes, and earthquakes to volcano eruptions—, no one-size-fits-all strategy for disaster preparedness emerges from these pages. Rather, by sharing a wide range of responses, this volume challenges the reader to think critically about disaster mitigation, including certain strategies that have become more popular in recent years, as well as their short- and long-term impacts. Among the most novel contributions in this regard are the chapters that focus on design competitions as well as on the shifting terminology deployed by humanitarian-shelter specialists.

One thread that does run through all the chapters is the focus on design, albeit deployed on a variety of scales. Given the urgency of housing in many disaster contexts, the contributing authors paid considerable attention to innovations in shelter typologies and the larger humanitarian discourse of architects. Yet, because disasters frequently require the reconstruction of social relationships that unfold on the neighborhood scale, and not merely the building scale, urban-design thinking is necessary to reconfigure commercial activities or community-based collective spaces. Equally important, the sensitivity to both building and urban-design thinking that permeates this volume is nestled within an appreciation for variations in the territorial scales of disaster response, from the local to the regional to the global.

In addition to what this volume offers, it is also worth noting what it astutely avoids. With a focus on disaster preparedness in its title, the editors do not cater to the obsession with resilience (although the word resilience does manage to appear in some of the chapter titles and section headings). Some might see this as an ill-considered move, if only because the notion of resilience has taken the disaster-related policy, design, and urban-planning worlds by storm. However, resilience is a tricky word, readily veering into the ideological. Defined as the ability to cope and adapt so that individuals or communities survive and thrive, resilience is all about *bouncing back to normal* after a disaster. Some authors use resilience to refer to the reestablishment of system equilibrium after a shock. Others use it as the rationale for a new and expanding repertoire of tools—from novel technologies to reconfigured mapping and building products—that guide us to a secure urban and global future. Whatever its application, those who emphasize resilience have faith that with enough attention and effort, the future can be better.

Yet, in addition to undervaluing if not ignoring the stark reality of accelerating risks, noted in the outset of this essay, many of those who embrace the concept of resilience tend to overlook the inter-relationalities of risk. There are trade-offs among forms and patterns of resilience, not just among different residents or between locations in the same city, but also in terms of immediate versus long-term gains in livability. Indeed, coping strategies in some domains (say environment) may actually reinforce structural problems that create risks in other domains (say inequality). Urban, social, economic, and environmental ecologies are connected locally and across scales that link cities to regions and beyond. Therefore, any resilience strategy must be grounded in an appreciation of the entire landscape of a city and its properties as a system embedded in a larger regional or even global ecology.

To the editors' credit, it is precisely these latter insights that thread through the chapters of this book and make it such a welcome addition to the disaster literature. These sensibilities are, for instance, well represented in the chapter that lays out a clustering methodology for land-use planning built around a nuanced understanding of ecology. Beyond individual contributions, a concern with inter-relationality across scales is also seen in the volume's overall organization, which moves from a focus on shelter in the first section to an examination of housing's embeddedness in community contexts in the second section. The third section carries forward the thread of housing and communities, but examines them through the lens of the global dynamics that keep many disaster-response agencies focused on shelter. The whole exercise lands with one of the volume's most synthetic pieces, built around a purposeful exploration of links between vulnerabilities, poverty, and disaster.

Given their professional backgrounds, this volume's editors know quite well that design thinking can be a tool for unpacking inter-relational complexities. Thanks to their practical experience, they are aware that adequate disaster preparedness is built around an understanding that any single design project or intervention will have implications far beyond its targeted scope, both in scalar and sectorial terms. Yet, to enable constructive action in a context of multiplying and interconnected vulnerabilities, it is also important to return to the idea of risk, and to design, build, and plan for a world of prevalent risk as much as for resilience. We must always stay prepared for the next disaster. This will require more than an ongoing engagement with new building and design techniques. Disaster experts will also need a new way of thinking about the connectivity of people, places, and spaces that allows communities to recover from one disaster while preparing for the next. Doing so successfully must involve interdisciplinary interaction and dialog among the various design, planning, and architecture professionals represented here, who will inevitably need allies in the social sciences, biological, engineering, technology, and public-health professions to prepare for a future of permanent risk. There is still much to be done, yet the pathway forward is already being charted, incrementally, through the grounded efforts and scholarly reflections contained in this timely volume.

Reference

Beck, U. (1992). *Risk society: Towards a new modernity*. Beverly Hills, CA: Sage.

Introduction

A. Nuno Martins[1], Mahmood Fayazi[2], Faten Kikano[3], Liliane Hobeica[4]

[1]CIAUD, Research Centre for Architecture, Urbanism and Design, Faculty of Architecture, University of Lisbon, Lisbon, Portugal; [2]The Institute for Disaster Management and Reconstruction, Sichuan University and the Hong Kong Polytechnic University, Chengdu, Sichuan, China; [3]Faculté de l'Aménagement, Université de Montréal, Montreal, Quebec, Canada; [4]RISKam (Research group on Environmental Hazard and Risk Assessment and Management), Centre for Geographical Studies, University of Lisbon, Lisbon, Portugal

In mid-August 2020, when we write these words, the world is going through an unprecedented emergency sanitary crisis. In about eight months after the first case was reported in China, the new Corona virus contaminated more than 21 million people, took more than 760 thousand lives, and caused the collapse of healthcare systems (WHO, 2020a). It has imposed the closure of borders and the lockdown of cities, as well as home quarantine and social-distancing practices on a scale never seen before. Further, the COVID-19 pandemic has severely affected national economies, decreased GDPs, and caused the loss of more than 300 million jobs worldwide. Given the absence of effective drugs, epidemiologists predict that the pandemic would continue for several months, if not years, until the development and implementation of a vaccine.

Like in other disaster situations, the pandemic has exacerbated regional and international inequalities. Although the sanitary crisis has affected individuals from all social groups, the World Health Organization considers that the most vulnerable populations in urban settings include dwellers of informal settlements, homeless persons, families living in inadequate housing conditions, forcibly displaced people, and migrants (WHO, 2020b). At this point, it is possible to anticipate the convergence of this pandemic with climate-related and human-induced crises in many geographies, which will eventually call for adequate cross-analyses. When vulnerability and poverty increasingly go hand in hand, and hazards shift from predicted patterns, extreme events should be taken as the new normal (UNDRR, 2019). The present context indeed highlights that preparedness should be duly entrenched in both regular development efforts and post-disaster settings, making the publication of this book even more pertinent.

Together with three other Elsevier books each related to one of the four priorities of the Sendai Framework for Disaster Risk Reduction (UNISDR, 2015), this publication is an output of the eighth edition of the International Conference on Building Resilience, held in Lisbon in November 2018. It gathers original contributions by authors from around the globe, most of them with a background in architecture and research. Nearly half of the chapters focus on humanitarian design whereas the others discuss community resilience. Together, they present a wide understanding of the fourth Sendai priority: "Enhancing disaster preparedness for effective response and to 'Build Back Better' in recovery, rehabilitation and reconstruction" (UNISDR, 2015, p. 14).

Mostly based on field research conducted in the Global South, this book deals with resilient responses and building capacities in relation to hazardous events, bringing some timely practical experiences and theoretical insights in this regard. It is organized in three parts. Part I, devoted to humanitarian architecture, puts together six contributions that address emergency sheltering and housing, disaster risk reduction (DRR), and post-disaster interventions (rebuilding and recovery). These contributions analyze communication and educational strategies with the aim of consolidating this field of knowledge. As a whole, they disclose the meaning and define the scope of humanitarian-architecture practice. The risk and resilience pair, as well as the notion of community design, permeates the assignment of architects engaged in both the disaster and development arenas. The changing role of architects and urban designers in times of climate change and the increasing number of vulnerable communities worldwide are thereby a bottom line to rethink architectural education and training.

Exploring humanitarian design and understanding resilience as a socioecological capacity that can be fostered through and within community-based DRR processes, Part II concentrates on humanitarian design and resilience building as a means to enhance community preparedness. In this regard, architecture, urban design, and community preparedness are addressed to face not only standalone disastrous events but also more regular urban threats and risks, such as eviction, gentrification, precarious housing, and health inequality. The four contributions in this part emphasize architects' diverse roles in supporting such capacity-building processes in the Global South, in either DRR, post-disaster, or development contexts. These roles eventually promote the full exercise of the "right to the city" (Lefebvre, 1968/1995), as stated in the comprehensive vision of the *New Urban Agenda* (UN-Habitat, 2017). As such, the authors call for architects and other built-environment professionals to not only foster the active participation of communities in DRR but also engage, with a spirit of consensus and compromise, in (aided) self-help design and construction processes. The chapters of Part III reveal that these processes can benefit from taking place in a framework that also acknowledges the responsibilities of governmental actors.

Part III brings new insights and fine nuances for concepts such as inclusive governance and community resilience. For instance, the authors here emphasize the link between global dynamics, whether economic or political, and national and local systems, with the vulnerability of communities. This awareness positively affects the adopted policies and responses in the management of crises at different levels of governance. Moreover, through a number of case studies, the four contributions in this part reveal novel aspects of community resilience. Strategies such as stakeholders' participation and the empowerment of communities affected by disasters, often depicted as key elements for DRR, are reassessed. The main messages of Part III imply that these approaches prove to be less beneficial if they are not concomitant with supervision by state officials, and guidance from experts.

In Chapter 1, Liz Brogden and Rosemary Kennedy delve into the inconsistencies and contradictions found in the humanitarian-shelter terminology. The authors consider that the plethora of terms, some of which being applied only in particular organizations or geographical contexts, inhibits the engagement of new practitioners and researchers in the humanitarian sphere. Thus, based on the review of 65 key documents, the authors developed a comprehensive shelter-terminology framework. The 8 categories and 25 subcategories gather 347 shelter terms, which concern both material and technical-support elements. Such a framework is a timely contribution to promote clearer understanding among stakeholders and the steady development of humanitarian architecture, planning, and engineering.

Judith I. Rodríguez Portieles's Chapter 2 presents a typical humanitarian-architecture experience. The author focuses on the joint recovery efforts of volunteers and community members in Puerto Rico in the aftermath of hurricanes Irma and Maria in 2017. She portrays the sheltering initiative undertaken by the Techo NGO, which, based on a participatory approach, supported affected communities by filling the gap of a deficient governmental response. Rodríguez Portieles highlights the role played by local architecture professionals to adapt a timber module to resist hurricanes and to meet the needs and preferences of beneficiary families. Her chapter pinpoints some best practices and areas of potential improvement in disaster recovery.

Eefje Hendriks, Benjamin Schep, and Alexander van Leersum, in Chapter 4, also cover a post-disaster reconstruction process, focused on the 2015 earthquakes in Nepal. Based on extensive fieldwork and resorting to social network analysis, their study sheds light on favorable conditions for the assimilation by local construction actors of knowledge on structural resistance to earthquakes. Comparing two districts that received dissimilar reconstruction technical assistance, the authors identify that communities in which external engineers had a major role displayed lower levels of understanding than those in which assistance provision involved a stronger network of local actors. This humanitarian-engineering study also emphasizes the need of increasing the dialog among the stakeholders in reconstruction processes.

In Chapter 3, Kristjana Adalgeirsdottir highlights lessons learned from the successful shelter response to a volcano eruption in Iceland in the 1970s. A national body then imported 479 prefabricated houses donated by the Nordic countries to fulfill the temporary housing needs. The author's detailed account demonstrates how the resilience of these temporary structures enabled them to remain in residential use even after their original users resettled back, in contrast with many other recovery cases worldwide (Lizarralde, Johnson, & Davidson, 2010). Through the new inhabitants' adaptations, extensions, and technical upgrading, these transitional houses succeeded to ultimately become permanent homes. The local management of the relief operations, the involvement of evacuees in decision-making processes, and the flexibility inherent to the structures were all factors that contributed to the endurance of the houses.

The theme of community involvement within humanitarian architecture is also recurrent in Chapter 5, which carefully details an experiment of participatory design led by Lara Alshawawreh in the Syrian refugee camps of Zaatari and Azrak in Jordan. The experiment aimed at identifying the refugees' shelter needs in terms of space, functions, and circulation. Refugees were solicited to design their "ideal" shelters by handling 3D mockups. Alshawawreh's findings reveal differences in the proposed designs according to the participants' gender, an often overlooked issue in shelter-provision operations. She also highlights the positive impacts of adapted built environments on refugees' wellbeing, especially those in protracted situations.

In Chapter 6, based on their pedagogical experiences and participation in the organization of design contests, A. Nuno Martins, Liliane Hobeica, Adib Hobeica, and Raquel Colacios analyze the 2018 and 2019 editions of an international humanitarian-architecture competition—the Building 4Humanity Design Competition (B4H-DC). To identify design patterns and explore previous successful experiences in bridging the gap between architectural education and disaster-recovery and reconstruction training, the authors also review the DRIA (Designing Resilience in Asia) and i-Rec (Information and Research for Reconstruction) international contests. Through an in-depth analysis of the B4H-DC winning projects, their research delves into the design tools employed by the competition participants to approach the involved design challenges, whether in DRR scenarios, post-disaster rebuilding and recovery, or forcedly displaced populations' settings.

Anouck Andriessen, Angeliki Paidakaki, Cynthia Susilo, and Pieter van den Broeck address in Chapter 7 the multiple roles played by architects in post-disaster settings to foster resilience, understood as a socially transformative capacity that supports bouncing forward in the aftermath of a shock. After conceptualizing these roles, the authors explore them in three reconstruction programs carried out following the 2010 eruption of the Merapi Volcano in Indonesia. They identify the governance structure and its institutional and programming rigidities as the major conditioning factors for architects' performance in reconstruction interventions. To overcome such limitations, the authors advocate that architects become more politicized to be able to exercise the full array of their competences in resilience-building processes.

In contrast, resilience is tackled in Chapter 8 from a prism other than that of disasters and conflicts. Helena Sandman and Miia Suomela explore design probing as a method to foster empathic engagement between communities and architects in processes of rapid and extensive spatial transformations in the Global South. Acknowledging the challenges that architects face when working with informal neighborhoods and the key roles of communities' active participation in supporting them to withstand shocks and also thrive, the authors present two examples of their own practice in Zanzibar, Tanzania. Their careful description and analysis of the two experiments show how design strategies can enhance communities' preparedness and enable them to deal with their daily struggles.

In Chapter 9, Georgia Cardosi, Susan Kibue, and Mauro Cossu also discuss the value of design in building resilience, this time considering the efforts of nonprofessionals in a setting characterized by the lack of tenure and urban facilities. The authors take an original standpoint to qualify as design actions the spatial transformations carried out by the traders of the Toi Market, one of the largest informal markets in Nairobi, Kenya. They argue that design thinking and ensuing practices have allowed the traders to adapt to and thrive between disturbing events, and illustrate how consolidation design has helped these slum dwellers to deal with risks, while strengthening their livelihood means.

Slums are likewise the focus of Chapter 10, in which Pablo Benetti and Solange Carvalho deal with the limits of government-led slum-upgrading processes and their DRR measures. Presenting several Brazilian cases of urban design in favelas, the authors claim that the lack of effective community participation and ownership, and the shortage in adequate maintenance by the municipal authorities have prevented the collective spaces ensuing from these initiatives from sustaining their status and conditions. For these projects to effectively meet DRR objectives, the authors propose the adoption of mechanisms that recognize the logic in favelas' expansion, giving voice and empowering local actors, and entrusting them with their role as the actual drivers of the urban development in their neighborhoods.

Participatory approaches do not comprise only positive outcomes. Drawbacks may occur when communities are left out without due information about disaster risks and proper design guidance regarding incremental-housing issues, even in formal settlements. These shortcomings are presented by David O'Brien and Sandra Carrasco in Chapter 12. The authors examine the case of the Villa Verde settlement in Chile—designed by Elemental— whose development coincided with the 2010 earthquake and tsunami that devastated the city of Constitucion. Without denying the benefits of the empowerment of residents who were encouraged to incrementally develop their houses according to their needs, the authors reveal that certain types of housing extensions adversely affected the settlement's livability and possibly increased wildfire risk. To avoid these downsides, they recommend a balance between participatory approach and collective governance in incremental-housing strategies.

In Chapter 13, Kelly Leilani Main, Miho Mazereeuw, Fadi Massoud, Jia Lu, Aditya Barve, Mayank Ojha, and Chetan Krishna propose an innovative response for adaptation to climate change consisting in building resilience through land-use planning rooted in eco-morphological attributes. Based on geospatial and flood-risk data, as well as clustering analysis, their experimental method entails the delineation of climatic action zones. These are then grouped into three categories—high-risk, low-risk, and uncertainty-oriented zones—each requiring particular management strategies whose governance extrapolates simple institutional boundaries. The clusters' environmental specificities are intended to guide future land-use planning and urban design, keeping in pace with rapidly changing ecological conditions, one of the most pressing dilemmas of our century.

In Chapter 11, Faten Kikano explores how global and local dynamics intertwine and impact on community resilience, focusing particularly on the case of refugees. Considering the protracted situation of Syrian refugees in Lebanon and based on extensive fieldwork, the author portrays the living conditions of these incomers and discusses some of the related drawbacks to the host country. Kikano claims that the policy of exclusion adopted by the Lebanese Government can nonetheless be altered in such a way as to benefit both refugees and local communities. She presents two key recommendations in this regard: a reorientation in the use of humanitarian funds and the temporary regularization of refugees' situation.

The interrelationship between local and global dynamics is further explored by Mahmood Fayazi and Lisa Bornstein in Chapter 14. Based on the review of a number of case studies, the authors first demonstrate the link between global trends and the economic vulnerability of societies. Then they skillfully identify the correlation between vulnerability to natural hazards and economic vulnerability and poverty. From a theoretical perspective, their findings highlight the link between different forms of vulnerability. Their conceptual approach can help practitioners and decision-makers, through their understanding of the multiple origins of vulnerability, in developing adapted solutions to mitigate the impacts of disasters on fragile communities.

Overall, the chapters in this book pinpoint to multiple interlinkages between humanitarian design, community resilience, and governance mechanisms regarding disaster preparedness, post-disaster rebuilding, and urban development. Despite their specificities, they share a few important take-home messages. One of these is that the most successful humanitarian-architecture and urban-design interventions are always capacity-building processes involving local communities and effective governance structures. Another lesson is that these processes benefit from balancing urban, architectural, social, and cultural dimensions. These recognitions shed light on the prominence of intrinsic human components in disasters and on growing vulnerabilities to poverty as well as to climate change in our increasingly unequal and unfair societies. Concomitantly, these lessons call for additional collective endeavors towards more equitable, safe, resilient, and climate-change adapted responses in our rapidly urbanizing world.

Acknowledgments

The editors would like to thank Adib Hobeica for his invaluable contributions to the review process.

References

Lefebvre, H. (1995/1968). Right to the city. In E. Kofman, & E. Lebas (Eds.), *Writings on cities* (pp. 61–181). London: Blackwell.

Lizarralde, G., Johnson, C., & Davidson, C. (Eds.). (2010). *Rebuilding after disasters: From emergency to sustainability.* Oxford: Spon Press.

UNDRR (United Nations Office for Disaster Risk Reduction). (2019). *Global Assessment Report on Disaster Risk Reduction.* Retrieved from https://gar.unisdr.org.

UN-Habitat. (2017). *New Urban Agenda.* Retrieved from http://habitat3.org.

UNISDR (United Nations International Strategy for Disaster Reduction). (2015). *Sendai Framework for Disaster Risk Reduction 2015–2030.* Retrieved from https://www.preventionweb.net/files/43291_sendaiframeworkfordrren. pdf.

WHO (World Health Organization). (2020a). *Coronavirus disease (COVID-19). Weekly epidemiological update 1.* Retrieved from https://www.who.int/docs/default-source/coronaviruse/situation-reports/20200817-weekly-epi-update-1.pdf?sfvrsn=b6d49a76_4

WHO (World Health Organization). (2020b). *Strengthening preparedness for COVID-19 in cities and other urban settings: Interim guidance for local authorities.* Geneva: WHO. Retrieved from https://apps.who.int/iris/handle/10665/331896

PART I

Humanitarian architecture

A humanitarian shelter terminology framework

Liz Brogden[1], Rosemary Kennedy[2]

[1]Queensland University of Technology, Brisbane, Queensland, Australia; [2]SubTropical Cities Consultancy, Brisbane, Queensland, Australia

1. Introduction

Organizations within an overwhelmed humanitarian system are increasingly turning to the private sector in search of collaborative partnerships to develop shelter solutions for displaced populations. However, the proliferation of shelter terminology and its inconsistent use in the shelter sector impedes development and obstruct new actors. Terminology influences the implementation of coherent sector principles and inconsistent use is a barrier to meaningful engagement from new partners seeking to access shelter sector knowledge. Further, misunderstood terminology limits the development of new strategic approaches and innovation. Zyck and Kent (2014, p. 18) highlighted that "exclusionary vocabularies" are evident across the humanitarian sector as an obstacle to collaboration. Bennett, Foley, and Pantuliano (2016) argued that certain terms and concepts represent a body of language that is only available to a small handful of Western universities who have a research focus on humanitarian affairs. This "retinue of anecdotes" (Bennett et al., 2016, p. 64) excludes and obscures access to knowledge and understanding of humanitarian shelter from those situated beyond the shelter sector itself.

There are calls to move away from a centralized and bureaucratic conception of the humanitarian system to one that is more open, flexible, and expansive. An open network of actors could accommodate new interpretations of what constitutes humanitarian action, as well as the recognition of new types of humanitarian actors. Bennett et al. (2016) described the humanitarian system as one that lacks a single, easily accessible entry point for new actors. This research explores patterns of shelter terminology, meaning, and use, which impede access to sector knowledge.

A *Google Images* search for "emergency shelter" returns an array of shelter prototypes, most of which are conceptual experiments developed in response to a humanitarian crisis that is rarely described. The images include expanding accordion-style structures, cocoons, teardrops, pods, tessellating hexagonal forms, and glowing "beacons of hope" shown in postapocalyptic landscapes (Google, 2018). It is uncommon to see *who* these shelters are intended to house, *where* they are to be located, for *how long* they will be occupied, or for *what type* of crisis. Well-meaning but misinformed design proposals are rarely grounded in the reality of a crisis or the needs of a community. Further, the sheer volume of information about humanitarian shelter that is dispersed across websites and databases worldwide is difficult to navigate, creating opportunities for "duplication, disagreement and inefficiency" (Knox Clarke & Campbell, 2015, p. 10).

This research explores a particular domain of humanitarian discourse emerging from 44 partners that make up the Global Shelter Cluster (2018a), as well as the Sphere Project (Sphere, 2018). The aim is to provide an overview of publications in the field, summarizing the array of shelter terminology in use across the shelter sector. A shelter terminology framework was developed using these terms, facilitating a common, systematic, and comprehensive understanding of shelter-specific terms and activities. Significantly, this research provides an interpretive tool to aid in accurately conceptualizing the problem of the shelter itself. This tool is intended to enable targeted engagement from practitioners and to facilitate research, practice, and education in this area. Additionally, the shelter terminology framework aims to contribute to the progression of the specialized fields of humanitarian architecture, planning, and engineering.

2. Literature review

It is often assumed that terms for shelter are implicitly understandable tacit knowledge. Further, publications relating to humanitarian shelter arise from academic institutions, private industry, and a vast number of organizations across the humanitarian sector. This body of knowledge reveals a wide range of terms that describe a comparably small variety of shelter types and approaches. Even when efforts are made to explain a shelter term, definitions are rarely interpretable beyond the context of a particular project or organization. For example, the glossary of terms in *Shelter after disaster: Strategies for transitional settlement and reconstruction* provides definitions for shelter and settlements, but these are the only terms that include the caveat: "For the purposes of these guidelines" (DFID & Shelter Centre, 2010, pp. 305—324).

Sector-wide, a cacophony of terms continues to multiply in an ever-growing number of reports. The 2011 *Sphere Handbook* outlined core humanitarian standards as "a practical expression of the shared beliefs and commitments of humanitarian agencies and the common principles, rights and duties governing humanitarian action" (Sphere, 2011a). Yet, the publication's "Minimum standards in shelter, settlements and non-food items" section reads as follows in the introduction chapter:

> Non-displaced disaster-affected populations should be assisted on the site of their original homes with temporary or transitional household shelter, or with resources for the repair or construction of appropriate shelter. Individual household shelter for such populations can be temporary or permanent, subject to factors including the extent of the assistance provided, land-use rights or ownership, the availability of essential services and the opportunities for upgrading and expanding the shelter [...] When such dispersed settlement is not possible, temporary communal settlement can be provided in planned or self-settled camps, along with temporary or transitional household shelter, or in suitable large public buildings used as collective centers *(Sphere, 2011a, p. 249).*

In one paragraph, six different terms are used to describe forms of shelter (original home, temporary shelter, transitional shelter, individual household shelter, transitional household shelter, and permanent shelter); two descriptors (upgradable and expandable); as well as five terms for settlement (dispersed settlement, temporary communal settlement, planned camp, self-settled camp, and collective center). Throughout the *Sphere Handbook*, additional shelter terms are introduced without definition, whereas transitional shelter is the only type defined in the handbook's text.

Furthermore, the accompanying online glossary for the *Sphere Handbook* omits any reference to shelter and its typologies altogether (Sphere, 2011b). After this research commenced, the 2018 version of the *Sphere Handbook* was released, which has a more consolidated approach in its use of shelter terminology. Definitions for shelter kits, shelter toolkits, tents, temporary shelters, transitional shelters, and core housing are provided in Appendix 4 of the document (Sphere, 2018, pp. 282—283).

The International Federation of Red Cross and Red Crescent Societies (IFRC), a lead partner in the Global Shelter Cluster, described the "overlapping definitions" shown in Fig. 1.1. The diagram is intended to identify and structure differing shelter terminologies (IFRC, 2013, p. 9), while also illustrating the associated qualities of each dwelling type (tent, module, simple house, established house).

According to the IFRC, shelter terms frequently relate to an "approach rather than a phase of response" (IFRC, 2013, p. 8). Terms for shelter in the IFRC's documents describe an overall process in which affected populations may build, upgrade, or maintain shelter in a way that changes its originally defined typology. The terms "progressive" and "incremental" have emerged to capture this phenomenon, yet the overlaps depicted in Fig. 1.1 arguably increase confusion rather than providing clarity for those unfamiliar with

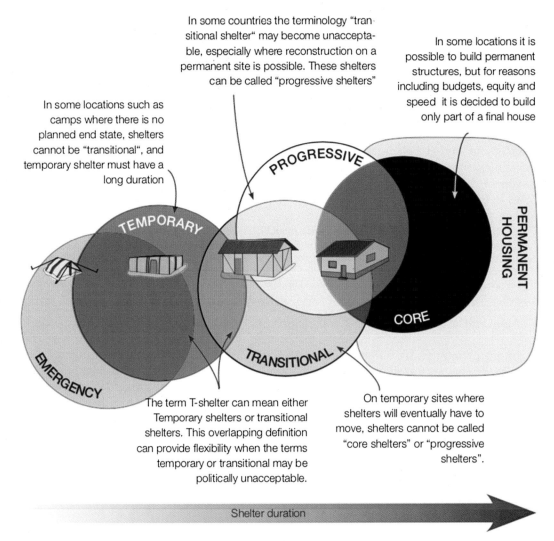

In some countries the terminology "transitional shelter" may become unacceptable, especially where reconstruction on a permanent site is possible. These shelters can be called "progressive shelters"

In some locations it is possible to build permanent structures, but for reasons including budgets, equity and speed it is decided to build only part of a final house

In some locations such as camps where there is no planned end state, shelters cannot be "transitional", and temporary shelter must have a long duration

PROGRESSIVE

TEMPORARY

PERMANENT HOUSING

EMERGENCY

CORE

TRANSITIONAL

The term T-shelter can mean either Temporary shelters or transitional shelters. This overlapping definition can provide flexibility when the terms temporary or transitional may be politically unacceptable.

On temporary sites where shelters will eventually have to move, shelters cannot be called "core shelters" or "progressive shelters".

Shelter duration

FIGURE 1.1 Overlapping definitions for shelter. *Reproduced with permission from IFRC. (2013). Post-disaster shelter: Ten designs. Retrieved from http://www.sheltercasestudies.org/files/tshelter-8designs/10designs2013/2013-10-28-Post-disaster-shelter-ten-designs-IFRC-lores.pdf. Geneva: IFRC, p. 9.*

shelter-sector activities. The IFRC does define shelter types, as revealed in the review carried out in *Shelter after disaster*, but the definitions provided are stated only "for the purposes of this book" (IFRC & UN-OCHA, 2015, p. 8). This is also the case for terms outlined in the 2018 *Sphere Handbook*, in which definitions cannot be applied confidently beyond the context of the document.

The intended meaning of shelter terms may be apparent to those embedded within the shelter sector, but caveats around shelter definitions reveal how complex the deciphering process can be for those who are less familiar. For example, to develop a project brief, an architect, engineer, or planner must fully understand the nature of the problem. A firm grasp of crucial terminology across a sector or industry is essential to an effective design or planning process and outcome. Further, in research, an understanding of key terms is also a fundamental component of a rigorously designed project.

2.1 A long-standing problem

Davis and Alexander (2016) stated that shelter has not been summarized adequately as the efforts of Davis (1978) in *Shelter after disaster*. Over two decades ago, Quarantelli (1995, p. 44) highlighted the problem of "multiple and ambiguous" meanings surrounding shelter terms, resulting in "contradictory baggages [sic] of connotations and denotations which do not allow for knowledge and understanding of the phenomena involved." Quarantelli (1995) observed that terms for the shelter came with the implicit assumption that they were self-explanatory. As such, he sought to define shelter terms according to four "ideal types": emergency sheltering, temporary sheltering, temporary housing, and permanent housing. In the years since, the number of shelter descriptors has continued to grow, often permutated with arbitrary interchanges between "shelter" and "housing."

Almost a decade later, during a review of the *Sphere Humanitarian Charter*, the issue of shelter terminology became more widely recognized in sector peer reviews and was identified as a significant obstacle to sector development (Shelter Centre, 2017). Saunders elaborated further observing that the absence of a common and coherent language for shelter and settlement weakened the shelter sector and was resulting in "major differences of opinion" (Saunders, 2004, p. 163). Saunders's discussion extended to question the name of the sector itself: "Is it shelter? Is it housing? Is it human settlements?" (Saunders, 2004, p. 161).

More recently, Boano and Hunter (2012, p. 3) referred to shelter and reconstruction practices in emergencies as reflecting a "profoundly semantic confusion," arguing that, when it comes to terms, "deciphering their nuances should be a necessity, as the consequences of conceptual confusion may create unwelcome results." The problem of unwelcome or inappropriate shelter has been observed by experts worldwide and is widely documented (Charlesworth, 2014; Davis, 2011; Duyne Barenstein, 2011; Fitrianto, 2011; Lizarralde, Johnson, & Davidson, 2010; Shaw, 2015). Boano and Hunter (2012) argued that conceptual confusion must first be removed to avoid poor shelter outcomes that are inappropriate to local conditions. They state that this deciphering is more than an academic exercise.

2.2 Terms describing phases of a shelter process

Most commonly, shelter terms are intended to be interpreted as part of a "three-stage recovery" model (Davis & Alexander, 2016). These stages begin with first-response emergency shelter, followed by medium-term temporary or transitional solutions, and finally permanent housing. Some experts have advocated for a two-stage model, removing the need for transitional shelter as a bridging phase between emergency shelter and permanent reconstruction, as seen in Fig. 1.2. The figure illustrates the three-stage versus two-stage conceptualization of shelter response, which is generally accepted in the shelter sector.

Not everyone agrees on a staged approach to reconstruction, though the terms employed to describe alternative approaches remain similar. The Shelter Centre's (2012) interpretation

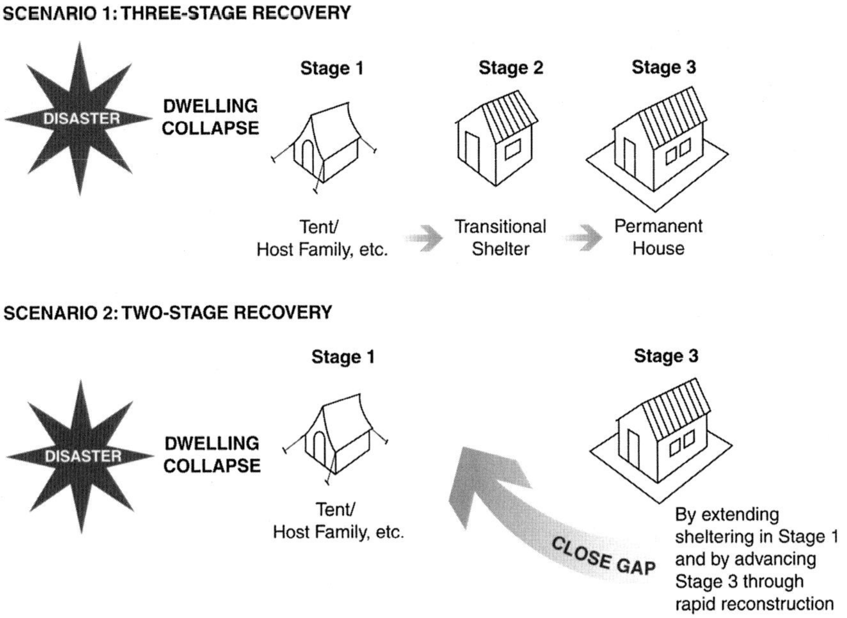

FIGURE 1.2 Three-stage and two-stage recovery scenarios. *Copyright 2016 From Recovery from disaster by David Alexander and Ian Davis. Reproduced by permission of Taylor and Francis Group, LLC, a division of Informa plc.*

of recovery distinguishes between incremental processes and a multiphased approach while advocating for incremental shelter. The Shelter Centre's ideal conceptualization is an incremental transition through a continuum from the immediate emergency to permanent housing. The multiphase model includes three discrete phases as shown in Davis and Alexander's three-stage recovery model, but the Shelter Centre uses the term "temporary shelter" rather than "transitional shelter" (see Fig. 1.3).

The *Transitional shelter guidelines* (Shelter Centre, 2012) sought to clarify shelter and settlement terms while also addressing issues of conceptual overlap. The publication provided answers to questions such as "are prefabricated shelters transitional shelters?" and "what is the difference between transitional shelter and core housing?" (Shelter Centre, 2012, p. 8). Despite this, the definitions provided in the publication have not been universally adopted, a fact evidenced by the continued proliferation of conflicting and contradictory shelter terms in the sector.

3. Research methods and materials

This study involved a systematic review of key publications and websites from the shelter sector. The focus was specifically on the Global Shelter Cluster (GSC), comprising organizations across all the major partners in the sector, including UN, Red Cross/Red Crescent, government, academia, and nongovernmental entities. The GSC was targeted for data collection to fulfill the objective of gaining a comprehensive sector-wide overview of language surrounding humanitarian shelter in both post-disaster and refugee response.

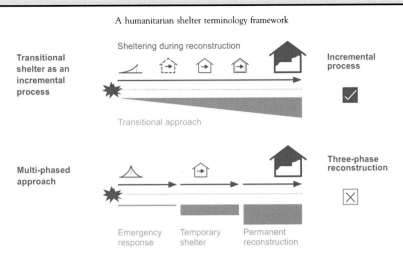

FIGURE 1.3   Transitional shelter as an incremental process. *Reproduced with permission from Shelter Centre. (2012). Transitional shelter guidelines. Retrieved from https://www.sheltercluster.org/resources/documents/transitional-shelter-guidelines. Geneva: Shelter Centre, p. 3.*

In all, 65 publications from the GSC and partner organizations were selected. In addition to these sources, the research included the three most recent *Shelter projects* publications (Global Shelter Cluster, 2013, 2015, 2017), the 2015 version of the *Sphere Humanitarian Charter* (Sphere, 2015), and three documents from each GSC partner member, as outlined in Table 1.1. The documents were sourced from web platforms, such as Humanitarian Response (UN-OCHA, 2017a), ReliefWeb (UN-OCHA, 2017b), and PreventionWeb (UNDRR, 2017), and included reports, online articles, websites, guidance materials, and research publications.

We applied qualitative content analysis as a method in view of its suitability for data in which manifest and latent meaning is context dependent (Elo & Kyngäs, 2008; Kohlbacher, 2006; Pickering, 2011; Schreier, 2013). A categorization technique was used to summarize, explain, and structure data (Kohlbacher, 2006). Schreier (2013) described the outcome of this method as a "coding frame" in which main categories and subcategories are generated and structured and then populated with references as encountered in the data, effectively defining those categories.

A list of shelter terms was built through the meta-analysis of the themes both in the sector and in the literature. Terms were then classified and mapped to a terminology framework and the main categories named. Through this process, areas of conceptual overlap or ambiguity within that framework were identified and further classified into subcategories. Manual coding was supported by *NVivo* software to efficiently search every instance of the word "shelter" and its associated synonyms: house, housing, and structure.

Many combinations of terms indicated a simple reordering or use of a synonym—for instance interchanging the operand "shelter," "house," or "structure." Frequently, this was seen to have no impact on the intended meaning, for instance, where "transitional shelter" and "transitional house" describe the same shelter strategy. However, these permutations sometimes impacted upon the intended meaning. An example of this is "durable shelter," a term that is appropriate for both first- and second-stage shelters, whereas "durable house" usually only describes a third-stage (permanent) solution. For this reason, we opted to include all the terms in the framework, despite the apparent repetition.

TABLE 1.1 Full list of the GSC partners' publications sourced.

Global Shelter Cluster partner's documents		
ACTED	2010	A shelter to recover
	2010	Shelter provision to flood-affected populations
	2014	Annual report
Australian Red Cross	2011	Gender and shelter
	2015	Annual review
	2016	Emergency shelter
British Red Cross	2011	Haiti one year on: From rubble to shelter
	2014	Trustees report and accounts
	2016	What is shelter?
CARE	2008	International policy brief on shelter
	2015	Emergency shelter team annual review
	2016	Post-disaster shelter in India: A study of the long-term outcomes of post-disaster shelter projects
Cordaid	2012	Final shelter report
	2015	2014 annual report
	2015	Shelter2Habitat: Developing resilient habitats after a disaster
CRS	2012	Learning from the urban transitional shelter response in Haiti
	2014	Annual report
	2016	Shelter and settlements
DFID	2011	Humanitarian emergency response review
	2015	Shelter from the storm by DFID UK: Exposure
	2015	Annual report 2014–2015
DRC	2015	2014 annual report
	2016	Danish Refugee Council provides emergency response after large scale destruction in Malakal, South Sudan
	2016	What we do
ECHO	2015	2014 annual report
	2016	Emergency shelter
	2016	Giving shelter
Emergency Architects Foundation	2014	Annual report 2013
Foundation		This foundation has offices in France and Canada; their Australian office is now closed. No English publications were available that met the selection criteria.

TABLE 1.1 Full list of the GSC partners' publications sourced.—cont'd

		Global Shelter Cluster partner's documents
German Red Cross	2011	Bangladesh: DRR in vulnerable communities
	2012	Disaster risk reduction in seven particularly vulnerable communities
	2015	Annual review 2014
Global Communities	2010	CHF builds pilot transitional shelter
	2014	2013 annual report
	2015	Better approaches needed for rapid rehousing after disasters
Habitat for humanity	2012	Disaster response shelter catalog
	2015	Annual report
	2016	Shelter report
IFRC	2012	Shelter lessons learned
	2013	Post-disaster shelter: Ten designs
	2014	Annual report
IMPACT		A working group of the GSC. No publication outputs found.
InterAction	2014	Annual report
	2015	Modules 1—5 notes, shelter and settlement training
	2016	Shelter
IOM	2013	Review of activities in disaster risk reduction and resilience
	2015	One room shelter: Building back stronger
	2015	Shelter highlights
IRC	2005	Shelter manual
	2013	Annual report
	2014	Growing humanitarian crisis in Iraq leaves thousands in need
Luxembourg Red Cross		Documents are in French.
Medair	2013	Medair expands shelter relief programme in the Philippines
	2014	Annual report
	2016	Shelter and infrastructure
NRC	2012	Urban shelter guidelines
	2013	Shelter
	2014	Annual report
OFDA	2012	Humanitarian shelter and settlements sector update
	2013	Humanitarian shelter and settlements principles
	2014	Description of humanitarian shelter and settlement activities
Oxford Brookes, CENDEP	2011	Good design in urban shelter after disaster
	2013	Changing approaches to post-disaster shelter
	2016	Shelter after disaster: Description of research area

(Continued)

TABLE 1.1 Full list of the GSC partners' publications sourced.—cont'd

Global Shelter Cluster partner's documents		
ProAct	2005	Emergency shelter environmental checklist
	2009	Environment training modules for emergency shelter
	2011	Annual review
Relief International	2014	Annual report
	2016	Haiti: Emergency and transitional shelter
	2016	Pakistan: Construction of 2000 temporary shelters for IDPs
Save the Children UK	n.d.	Protection of children in emergency shelters
	2014	Annual report
	2015	Iraq: No windows, no roof. But for now, this is home
Shelter Centre	2010	Annual report 2009–2010
	2012	Transitional shelter guidelines
Shelter for Life International	2006	Transitional shelter assistance in Tajikistan
	2014	Annual report
	2016	What we do: Shelter
ShelterBox	2014	Annual report
	2016	Deadly earthquake strikes in Ecuador
	2016	Desperate need for shelter in Fiji in the wake of cyclone Winston
Swedish Red Cross	2012	Annual report
		See publications for IFRC.
UN-Habitat	2005	Financing urban shelter: Global report on human settlements
	2011	Enabling shelter strategies
	2013	Global activities report
UNHCR	2014	Global report
	2014	Global strategy for settlement and shelter
	2016	What we do: Shelter
UN-OCHA	2005	Humanitarian response review
	2015	Shelter after disaster
	2016	Key things to know about the emergency shelter cluster
World Vision International	2012	Minimum inter-agency standards for protection mainstreaming
	2014	Annual report
	2016	Shelter and warm clothes for El Niño affected people

As terms were encountered in the data, the intended contextual or latent meaning was interpreted pragmatically. The shelter terminology framework (or coding frame) was maintained as dynamic and adjustable throughout the coding process.

4. The shelter terminology framework

Data saturation was reached at 347 terms describing shelter. The final framework consists of eight main categories: immediate shelter; intermediate shelter; permanent shelter; preemptive shelter; nonspecific shelter terms; shelter items; alternative strategies; and multiphase shelter. Each category is explained by two-to-four subcategories, resulting in 25 ways to describe shelter strategies, stages, types, and artifacts. The full shelter terminology framework is shown in Fig. 1.4.

In constructing the shelter terminology framework, we identified that the term "transitional shelter" is a significant source of terminological confusion. As indicated in Fig. 1.4, two

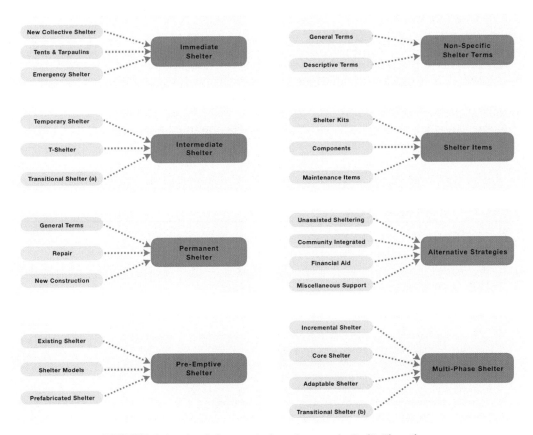

FIGURE 1.4 The shelter terminology framework. *Credit: The authors.*

subcategories of transitional shelters are represented. The first is in the "second-stage shelter" category (type a), and the second in the "multiphase shelter" category (type b). These reflect a distinction that is discussed extensively in the *Transitional shelter guidelines* (Shelter Centre, 2012) and also the updated edition of *Shelter after disaster* (IFRC & UN-OCHA, 2015). The subcategory "transitional shelter (a)" denotes a designed product intended to serve a purpose within a discreet second reconstruction phase. The subcategory "transitional shelter (b)" describes an incremental process.

4.1 Immediate shelter

This category summarizes terms associated with the first response (see Table 1.2). Shelter terms incorporated in this category refer to strategies in which minimal response time is the priority to provide life-saving assistance. These shelter strategies are usually intended only to serve a short-term purpose. This category includes new collective shelter centers, lightweight tents and tarpaulins, and basic structural materials that are frequently also called as emergency shelters.

4.2 Intermediate shelter

This category incorporates semipermanent shelter strategies intended to serve a medium-term purpose, or an average of three to five years (see Table 1.3). It includes the subcategories: temporary shelter, product-focused notions of transitional shelter, and some types of transportable shelter. All of these can be described as "T-shelters" depending on the context. These "T-words" are often used interchangeably to describe intermediate shelters by GSC partners, even within the same document.

TABLE 1.2 Category 1—Immediate shelter.

Total	Terms	Subcategory	Category
3	Collective center; transit centers; return centers	New collective shelter	
13	Family tent; government-supplied tent; makeshift tent; phantom tent; plastic sheeting; shelter tent; tarp; tarpaulin sheeting; tarp-clad shelter; tent structure; tent on concrete; emergency tent; shelter-grade plastic sheeting	Tents and tarpaulins	1 Immediate shelter
20	Emergency shelter; early shelter; emergency housing; emergency shelter (temporary); emergency structure; immediate shelter; initial shelter; phase-one shelter; rapid shelter; short-term emergency shelter; temporary emergency shelter; urgent shelter; emergency family shelter; recovery shelter; emergency-shelter kit; special emergency shelter; early recovery shelter; immediate emergency-response shelter; rapidly deployable shelter; short-term shelter	Emergency shelter	

TABLE 1.3 Category 2—Intermediate shelter.

Total	Terms	Subcategory	Category
9	Temporary shelter; temporary accommodation; temporary home; temporary house; temporary resettlement site; temporary structure; temporary emergency shelter; shelter (temporary); temporary-shelter kit	Temporary shelter	
7	Two-storey T-shelter; rural T-shelter; urban T-shelter; transitional shelter (T-shelter); T-shelter phase; transitional shelter (T-shelter); T-shelter kit	T-shelter	2 Intermediate shelter
12	Transitional home; transitional house; transitional shelter (semipermanent); transitional shelter (T-shelter); upgraded transitional shelter; urban transitional shelter; expandable transitional housing; transitional night shelter; semipermanent transitional shelter; transitional-shelter kit; transitional individual household shelter; transitional household shelter	Transitional shelter product (a)	

TABLE 1.4 Category 3—Permanent shelter.

Total	Terms	Subcategory	Category
18	Permanent shelter; final house; lifetime houses; long-term housing; long-term shelter; durable structure; durable building; integrated housing; permanent durable house; permanent durable shelter; permanent home; permanent house; durable home; durable house; permanent structure; post-disaster house; durable solution; long-term shelter	General terms (third stage)	
5	Repaired housing; as-built shelter; shelter in-kind; rehabilitated shelters; repaired dwelling	Repair	3 Permanent shelter
8	Concrete house; contractor-built houses; concrete shelter; permanent core house; new house construction; permanent reconstruction housing; permanent reconstruction; mud-brick shelter	New construction	

4.3 Permanent shelter

This category denotes shelter strategies associated with long-term recovery (see Table 1.4). These include descriptive words such as durable and concrete shelter. Three subcategories are identified within this category: general terms describing a permanent shelter outcome, terms associated with the repair of damaged housing, and terms indicating new construction or reconstruction.

4.4 Preemptive shelter

This category outlines shelter terms describing strategies conceived before a crisis event (see Table 1.5). Subcategories include existing shelter, shelter models, and prefabricated shelter. Existing shelters include locations in the community intended as evacuation points such as cyclone shelters, and also existing housing. Terms in the shelter model's subcategory

TABLE 1.5 Category 4—Preemptive shelter.

Total	Terms	Subcategory	Category
13	Predisaster shelter; community spaces; institutional structure; cyclone shelter; evacuation shelter; existing housing; pre-disaster home/house; safe shelters; school evacuation center; collective center; precrisis shelter; existing public building	Existing shelter	
13	Shelter prototype; core-house model; demonstration shelter; demonstration unit; model house; pilot emergency shelter; pilot transitional shelter; temporary shelter prototype; T-shelter model; trial shelter; pilot shelter; model shelter; prototypical shelter design	Shelter models	4 Preemptive shelter
20	Prefabricated shelter; deployable shelter; fabrication unit; modular housing; modular shelter; precast T-shelter; prefab transitional housing; shelter module; imported shelter; transportable shelter; transportable housing; shelter stockpile; prefabricated shelter unit; prefab unit; prefab shelter; premanufactured portable shelter; prefabricated container shelter; moveable shelter; container shelter; foldable shelter; modular light shelter	Prefabricated shelter	

TABLE 1.6 Category 5—Nonspecific shelter terms.

Total	Terms	Subcategory	Category
25	Developmental housing; disaster-response shelter; essential shelter; housing unit; humanitarian shelter; post-disaster shelter; post-disaster structure; post-earthquake house; primary shelter; shelter after disaster; shelter product; shelter facilities; shelter unit; shelter solution; shelter strategy; shelter structure; small shelter unit; nonfood item (NFI); relief shelter; postcrisis shelter; individual shelter; individual housing; shelter assistance; shelter intervention; basic NFIs	General terms (nonspecific)	
38	Standardized shelter; disaster-friendly houses; disaster-mitigation housing; disaster-resilient housing; disaster-resilient shelter; disaster-resilient structure; earthquake-recovery housing; earthquake-resistant housing; one-room shelter; basic shelter; two-room shelter (TRS); intermediate shelter; semipermanent shelter; semi-permanent home/house; medium/long-term shelter; impermanent shelter; reinforced shelter; scaled shelter solution; short-term house; hybrid structure; night shelter; modular tunnel shelter; noncamp shelter; flood-resilient shelter; refugee shelter; accessible shelter; bathing shelter; IDP shelter; participatory shelter; covered living space; covered space; covered area; dome shelter; mud house; hut; refugee housing unit	Descriptive terms (nonspecific)	5 Nonspecific shelter terms

describe demonstration, or pilot products and prototypes that are not necessarily intended for immediate use or deployment. Finally, the prefabricated shelter subcategory contains terms relating to the shelter that is mass manufactured in anticipation of a crisis event. This subcategory may also describe products suited to other categories (for instance immediate, intermediate, or permanent shelter solutions), however, due to the preemptive nature of a prefabricated-shelter strategy, we included it as a discrete shelter subcategory.

4.5 Nonspecific shelter terms

This category incorporates all the terms that refer to shelter in a way that is either general or descriptive, without denoting a specific shelter type (see Table 1.6). General terms describe humanitarian shelter broadly as a phenomenon, for example: "post-disaster shelter," "disaster response shelter," and "relief shelter." Descriptive terms explain a quality of a particular shelter but are not intended to refer to a commonly understood shelter strategy or type, for example: "disaster-resilient shelter," "reinforced shelter," and "basic shelter."

4.6 Shelter items

"Shelter items" relate to kits, components, or maintenance items (see Table 1.7). It is not always clear whether kits refer to a complete shelter solution or not, but this subcategory generally refers to a strategy in which beneficiaries are required to construct the shelter themselves. Shelter kits may serve both immediate and intermediate shelter needs. Terms in the components subcategory suggest a less comprehensive assistance package, in which additional items may be required that are not included. Maintenance items include any ongoing requirements for repair and materials for the seasonal adaptation of shelters.

4.7 Alternative strategies

This category refers to shelter strategies that do not involve material assistance (see Table 1.8). It incorporates terms describing unassisted (self-built) shelter solutions, community-integrated, financial-aid programs, and miscellaneous support for capacity building and advocacy.

TABLE 1.7 Category 6—Shelter items.

Total	Terms	Subcategory	Category
23	Covering kit; emergency-shelter kit; essential-items kit; household-shelter kits; T-shelter kit; temporary-shelter kit; transitional-shelter kit; return-and-repatriation kit; shelter-covering kit; shelter kit; shelter-repair kit; emergency-shelter kit; standardized-shelter kit; recovery-shelter kit; NFI kit; mobile NFI kit; mobile emergency-shelter kit; in-kind kit; full sealing-off kit; family-shelter kit; shelter-relief kit; shelter-recovery kit; portable-shelter kit	Shelter kits	6 Shelter items
18	Key in-kind shelter items; shelter components; emergency-shelter materials; construction-materials kits; shelter toolkit; key shelter items; shelter materials; community toolkits; material package; material vouchers; household nonfood items; shelter nonfood items; basic building materials; tools; relief items; prefabricated parts; shed nets; insulation	Components	
10	Minor shelter-repair kit; major shelter-repair kit; sealing-off kit; winterization package; seasonal shelter items; emergency sealing-off kit; climatization package; shelter-fixing kit; shelter toolkit; shelter-repair kit	Maintenance items	

TABLE 1.8 Category 7—Alternative strategies.

Total	Terms	Subcategory	Category
11	Vernacular shelter; shelter self-recovery; self-made shelter; self-built shelter; salvaged-materials shelter; rudimentary shelter; makeshift shelter; self-sheltering solution; debris-to-shelter approach; community labor; urban self-settlement; rural self-settlement	Unassisted sheltering	
10	Community shelter; half-way house; host community; host family; rental accommodation; subdivided houses; short-term flat; hosting; host shelter; host/ing support	Community integrated	
16	Cash-assistance program; cash grant; emergency cash grant; rental-subsidies program; rental support; resettlement grant; shelter-subsidy scheme; loan voucher; rental assistance; rental arrangement; conditional cash grant; cash-transfer program; cash-for-rent; cash-based transfer; vouchers; loans and guarantees	Financial aid	7 Alternative strategies
22	Structural assessment; training; site planning; shelter workshops ; nonmaterial assistance; legal support; advocacy; guidelines assistance; construction training; capacity building; promoting livelihoods; technical training; contracted labor; direct labor; local information centers; technical expertise; program integration; market support; return-and-transit items; infrastructure; environmental management; specialist labor	Miscellaneous support	

TABLE 1.9 Category 8—Multiphase shelter.

Total	Terms	Subcategory	Category
8	Progressive shelter; incremental housing; upgradable transitional housing; upgraded transitional shelter; upgradable shelter; upgradable house; enhanced emergency shelter; incremental building	Incremental shelter	
8	Permanent core house; core building; core house; core shelter; core structure; part (core) shelters; shelter core; single-room core shelter	Core shelter	8 Multiphase shelter
15	Flexible shelter; extendable shelter; expandable shelter; scalable shelter; flexible shelter; retrofitting/reuse/recycling; expandable housing; transferable shelter; enhanced shelter; reusable shelter; resaleable shelter; collapsible shelter; resellable shelter	Adaptable shelter	
2	Shelter continuum; shelter process—also see transitional shelter (a)	Transitional-shelter process (b)	

4.8 Multiphase shelter

This category is indicative of a move toward process-based thinking in the shelter sector (see Table 1.9). Rather than developing discrete products, this category outlines strategies intended to flank multiple stages of response. These strategies include incremental-shelter processes in which a shelter can evolve to become permanent, core shelter products to support similar incremental improvements, various adaptable solutions, and process-based notions of transitional shelter.

5. Discussion

The shelter terminology framework addresses a significant problem impeding participation in the humanitarian sector by potential new actors. The veracity of interpretation regarding shelter terminology is fundamental to ensuring positive outcomes for people impacted by the significant upheaval of disasters or conflicts. This research pays particular attention to those efforts made by GSC member groups and the network of connected terms informing the shelter discourse. It is this network—or framework—that is presented here as a tool for conceptualizing interconnected shelter phases, products, and processes, rather than a focus on individual shelter definitions. The shelter terminology framework expands upon a three-phase model of shelter response, incorporating notions of immediate, intermediate, and permanent phases while acknowledging the overlap of terms in particular contexts that are intended to describe a multiphase shelter process. The framework identifies terms that denote preconceived (preemptive) shelter strategies, which are intended to serve a purpose across one or more shelter phases, dependent on the context. Additionally, the shelter terminology framework acknowledges nonmaterial shelter support and bottom-up initiatives.

The framework derived out of this research serves to structure and categorize shelter terms as encountered in the data. It is designed to be a tool that can aid in conceptualizing shelter strategies by illustrating patterns in terminology use. It is intended to be flexible and adjustable and to provide a shared conceptual language. It is not presented as exhaustive, nor is it an authoritative list of shelter types or definitions. We assert that the primary value of the framework lies in its use in assisting in deciphering nuances and facilitating the understanding of shelter terms in context. Additionally, the overall framework and the descriptions of categories impart a condensed overview of shelter types and activities that is readily graspable by existing and new actors alike.

One explanation for terminological ambiguity could lie in the frequent interchangeability of "T" words such as transitional, temporary, transportable, and "T-shelter." The *Transitional shelter guidelines* define transitional shelter as "an incremental process which supports the shelter of families affected by conflicts and disasters, as they seek to maintain alternative options for their recovery" (Shelter Centre, 2012, p. 2). The guidelines also outlined five characteristics of transitional shelter, stating that it can be: "(1) upgraded into part of a permanent house; (2) reused for another purpose; (3) relocated from a temporary site to a permanent location; (4) resold, to generate income to aid with recovery; and (5) recycled for reconstruction" (Shelter Centre, 2012, p. 2).

The definition of transitional shelter is moving to encompass all shelter types except those that are a final shelter built at once, or any prefabricated shelter. The guidelines also exclude core housing from its definition of transitional shelter. Core shelter is the construction of part of a shelter, onto which the permanent house can be built upon, whereas the guidelines specify that "the approach does not allow reuse for another purpose, the relocation to another site or recycling of components for permanent reconstruction" (Shelter Centre, 2012, p. 8). This distinction could be problematic for those unfamiliar with nuances of shelter types, especially given that the first characteristic listed in the guidelines includes instances where shelter is upgraded into a permanent house. It is unclear why reuse, relocation, recycling, or reselling are stated as determining factors in whether a core shelter is deemed to be "transitional," given the guidelines' incorporation of incremental shelter processes within the definition.

Conceptualizing transitional shelter as multiphase or incremental points to the potential of design thinking in developing innovative shelter solutions. The Global Shelter Cluster (2017, p. x) states that "shelter is 'more than just a roof', it is not just the structure that protects from the elements, but is the series of activities that a household undertakes to save and construct, adapt and expand a dwelling, as well as the range of continuing actions and livelihoods that people do in and around their home." For architects, responding to social dimensions in the creation of space is fundamental to good design practice. As is conceiving buildings situated within a broader context of complex processes that extend to an urban, and even global scale. The notion of transitional shelter as a process rather than a product speaks powerfully to the potential of architecture and design as the best point of engagement for these professions.

Discourse about shelter in the humanitarian sector is moving toward its conceptualization as a process and the incorporation of modes of shelter assistance that are support-based. This trend is reflected in the shelter terminology framework, which includes terms that describe both material and nonmaterial shelter strategies. Although most of the terms relate to material assistance—as reflected in seven of the eight main categories—, nonmaterial shelter strategies are included within the "alternative strategies" category. This category could be seen as paradoxical in that there are shelter strategies that are characterized as nonmaterial, but also terms that describe the total absence of any external support (terms relating to self-sheltering activities). In many cases, the best shelter support is capacity building to enable self-recovery.

Nonmaterial approaches include advocacy, expert assessments, community consultations, cash support, or community-resettlement programs. We deemed it appropriate to include self-sheltering activities in the "Alternative strategies" category to acknowledge scenarios in which the decision to "leave well alone" is the most appropriate course of action. Particularly in situations in which damage to housing is assessed as minimal, or if a community has the resources required to recover from a disaster. We assert that this interpretation is in contrast to unassisted sheltering, when external material aid is required but not provided. This situation generally arises when needs exceed the response capacity of humanitarian agencies or governments due to the increasing scale and complexity of disasters (Davis & Alexander, 2016).

It must be noted that the presented shelter terminology framework does not account for socioeconomic factors such as tenure, nor residential building types and urban typologies such as row houses and apartments. There are also no categories that enable the distinction between context-specific shelter strategies particular to urban, peri-urban, or rural settings. Furthermore, as stated in the previous section, this research does not delve into inconsistencies that exist around terms for settlements, nor the humanitarian-shelter sector itself.

This research does not explore the impact of geographic variation of terms identified by Saunders (2004) as a source of contradiction and confusion, nor does it examine the impact of language and the effects of translation on shelter terminology. Furthermore, the framework reflects a Western interpretation of housing and humanitarian aid, and new actors must be cognizant of further overlays and local nuances, including local knowledge and resources.

6. Conclusions

The developed shelter terminology framework is expected to progress humanitarian fields and emergent specialist areas of design and planning, by enabling targeted engagement from practitioners through both research and education. The framework illustrates patterns of use and the variations in understanding surrounding terms for other shelters and settlements and makes sense of a complex area of human endeavor.

The significance of this research for practice and across the shelter sector is in enabling greater clarity and alignment between different clusters, an objective set in the GSC's 2018–2022 shelter and settlements strategy (Global Shelter Cluster, 2018b). The strategy's narrative identifies that integrated, localized responses are critical for successful shelter projects, starting with an "agreement on definitions and understanding of terminology" (Global Shelter Cluster, 2018b, p. 19). Furthermore, it is anticipated that in promoting clarity, the provided shelter terminology framework might also reduce the difficulties associated with the accurate interpretation of shelter-sector information and prevent what Boano and Hunter (2012, p. 3) describe as "unwelcome results" seen in the shelter sector.

For built-environment professions, it is anticipated that architects, engineers, and planners equipped with a clear understanding of the nature of humanitarian-shelter types and approaches are more likely to contribute meaningfully. Additionally, in education, enabling clear and consistent interpretations of shelter terminology will impact positively on research and courses recently emerging from academic institutions.

For shelter research, this study holds significance for a multitude of specialist fields that are emerging in the private sector as a result of intensified partnerships with the shelter sector. These include various industries of development, finance, business, technology, and design. Looking ahead, this research illustrated how the evolution of the term and nature of "transitional shelter" highlights a nexus point where fundamental changes in thinking about shelter and the future of the shelter sector lie, particularly in the understanding of shelter as a process and not simply a product or unit. Additionally, the need to conceptualize shelter as integrated within broader settlement systems is essential in developing successful humanitarian-shelter strategies.

References

Bennett, C., Foley, M., & Pantuliano, S. (2016). *Time to let go: Remaking humanitarian action for the modern era*. Retrieved from https://www.odi.org/publications/10381-time-let-go-remaking-humanitarian-action-modern-era.

Boano, C., & Hunter, W. (2012). Architecture at risk (?): The ambivalent nature of post-disaster practice. *Architectonica, 1*(1), 1–13.

Charlesworth, E. (2014). *Humanitarian architecture: 15 stories of architects working after disaster*. New York, NY: Routledge.

Davis, I. (1978). *Shelter after disaster*. Oxford: Oxford Polytechnic Press.

Davis, I. (2011). What have we learned from 40 years' experience of disaster shelter? *Environmental Hazards, 10*(3–4), 193–212. https://doi.org/10.1080/17477891.2011.597499.

Davis, I., & Alexander, D. (2016). *Recovery from disaster*. Abingdon, Oxon: Routledge.

DFID, & Shelter Centre. (2010). *Shelter after disaster: Strategies for transitional settlement and reconstruction*. Retrieved from https://reliefweb.int/report/world/shelter-after-disaster-strategies-transitional-settlement-and-reconstruction.

Duyne Barenstein, J. E. (2011). The home as the world: Tamil Nadu. In M. Aquilino (Ed.), *Beyond shelter: Architecture for crisis* (pp. 184−195). London: Thames and Hudson.

Elo, S., & Kyngäs, H. (2008). The qualitative content analysis process. *Journal of Advanced Nursing, 62*(1), 107−115.

Fitrianto, A. (2011). Learning from Aceh. In M. Aquilino (Ed.), *Beyond shelter: Architecture for crisis* (pp. 26−39). London: Thames and Hudson.

Global Shelter Cluster. (2013). *Shelter projects 2011−2012.* Retrieved from https://www.sheltercluster.org.

Global Shelter Cluster. (2015). *Shelter projects 2013−2014.* Retrieved from https://www.sheltercluster.org.

Global Shelter Cluster. (2017). *Shelter projects 2015−2016.* Retrieved from https://www.sheltercluster.org.

Global Shelter Cluster. (2018a). *About us.* Retrieved from http://www.sheltercluster.org/working-group/about-us.

Global Shelter Cluster. (2018b). *Global Shelter Cluster strategy 2018−2022 − narrative.* Retrieved from https://www.sheltercluster.org.

Google. (2018). *Google images search: "Emergency shelter".* Retrieved from http://www.google.com.

IFRC. (2013). *Post-disaster shelter: Ten designs.* Retrieved from http://www.sheltercasestudies.org/files/tshelter-8designs/10designs2013/2013-10-28-Post-disaster-shelter-ten-designs-IFRC-lores.pdf.

IFRC, & UN-OCHA. (2015). *Shelter after disaster* (2nd ed.). Geneva: UN-OCHA, IFRC.

Knox Clarke, P., & Campbell, L. (2015). *Exploring coordination in humanitarian clusters.* Retrieved from https://www.alnap.org/help-library/exploring-coordination-in-humanitarian-clusters.

Kohlbacher, F. (2006). The use of qualitative content analysis in case study research. *Forum: Qualitative Social Research, 7*(1). Retrieved from http://www.qualitative-research.net/fqstexte/1-06/06-1-21-e.htm.

Lizarralde, G., Johnson, C., & Davidson, C. (2010). *Rebuilding after disasters: From emergency to sustainability.* New York, NY: Taylor and Francis.

Pickering, M. J. (2011). Qualitative content analysis. In M. S. Lewis-Beck, A. Bryman, & T. F. Liao (Eds.), *The Sage Encyclopedia of Social Science Research Methods.* Thousand Oaks, CA: Sage.

Quarantelli, E. L. (1995). Patterns of sheltering and housing in US disasters. *Disaster Prevention and Management, 4*(3), 43−53.

Saunders, G. (2004). Dilemmas and challenges for the shelter sector: Lessons learned from the Sphere revision process. *Disasters, 28*(2), 160−175.

Schreier, M. (2013). Qualitative content analysis. In U. Flick (Ed.), *The Sage Handbook of Qualitative Data Analysis* (pp. 170−183). London: Sage.

Shaw, R. (2015). *Recovery from the Indian Ocean tsunami: A ten-year journey.* Tokyo: Springer.

Shelter Centre. (2012). *Transitional shelter guidelines.* Retrieved from https://www.sheltercluster.org/resources/documents/transitional-shelter-guidelines.

Shelter Centre. (2017). *Humanitarian library: The global hub for humanitarian and development knowledge.* Retrieved from http://www.humanitarianlibrary.org.

Sphere. (2018). *The Sphere Handbook: Humanitarian Charter and minimum standards in humanitarian response* (2018 ed.) Retrieved from https://spherestandards.org/handbook/editions.

Sphere. (2011a). *Humanitarian Charter and minimum standards in humanitarian response.* Retrieved from https://spherestandards.org/handbook-2018.

Sphere. (2011b). *Sphere Handbook 2011 edition glossary.* Retrieved from http://www.sphereproject.org/handbook/glossary.

Sphere. (2015). *The Sphere Project: Humanitarian Charter and minimum standards in humanitarian response.* Retrieved from http://www.sphereproject.org/about.

UN-OCHA. (2017a). *Humanitarian response: Document database.* Retrieved from https://www.humanitarianresponse.info.

UN-OCHA. (2017b). *ReliefWeb International: Informing humanitarians worldwide.* Retrieved from http://reliefweb.int.

UNDRR. (2017). *PreventionWeb knowledge base. Documents and publications.* Retrieved from https://www.preventionweb.net/english/professional.

Zyck, S. A., & Kent, R. (2014). *Humanitarian crises, emergency preparedness and response: The role of business and the private sector.* Retrieved from https://www.odi.org/publications/8534-humanitarian-crises-emergency-preparedness-and-response-role-business-and-private-sector-final.

Techo's emergency-housing response to hurricanes in Puerto Rico: Lessons from the field

Judith I. Rodríguez Portieles

Department of Architecture, Harvard University Graduate School of Design, Cambridge, MA,
United States

1. Introduction

The loss of a home in a disaster is a traumatic experience from the individual, social, and economic points of view. The provision of temporary or transitional housing is thus a crucial component of disaster recovery, as it allows people to return to normalcy while ensuring protection, privacy, and dignity (Félix, Branco, & Feio, 2013; IFRC, 2011). Yet, many rebuilding efforts take housing as the supply of a physical product by procedures exogenous to the community, rather than viewing it more deeply as a demand-driven process of reconstituting homes instead of mere houses (Sanderson, 2018). According to Davis (2011), the adoption of a process perspective can be challenging for shelter professionals and manufacturers, who tend to tackle shelter provision through simple problem-solving approaches.

The quality of temporary housing is influenced by its design, cultural appropriateness, timeliness, location, implementation process, and the extent to which it involves the affected families—aspects that impinge on their recovery capacity (Félix et al., 2013). Prefabricated and mass-produced modules that are hard to customize permeate many of the top-down approaches to temporary housing (Félix et al., 2013). These types of modules respond to the speed and efficiency required in emergency responses, but entail fewer interactions, if any, with the end-users. Moreover, the deployment of undifferentiated ready-made units, to compose homogeneous settings often located far from a community's previous settlement, eventually disconnects families from their social networks.

In contrast, housing modules in bottom-up transitional-housing responses allow for customization through the engagement of the affected communities. Kit-based solutions, composed of prefabricated parts to be assembled at the site, often require the participation of local people (Félix et al., 2013), therefore fostering more engagement with the beneficiary communities. This strategy seems thus more aligned with current arguments of shelter practitioners and scholars alike, highlighting "the primary role of disaster survivors in providing their own shelter and the need for participation of beneficiaries at every stage in their recovery" (Davis, 2011, p. 210).

According to Comerio (2014), disaster recovery through a people-focused approach has the advantage of helping to empower affected families into active stakeholders, who can eventually participate more effectively in the decisions related to their future. The provision of temporary housing can be embedded in people-focused processes. A well-integrated shelter initiative can help to fulfill the social, emotional, and psychological needs of disaster survivors by providing them with additional security, as well as therapy and solidarity from working together (Davis, 2011).

This chapter examines an emergency-housing response in a low-income Puerto Rican community affected in 2017 by the hurricanes Irma and Maria. The study aimed to understand how a participatory volunteer initiative could assist the affected community in regaining control over their circumstances and improving their quality of life through shelter provision. I reviewed the transitional-shelter experience led by the Techo NGO with volunteers through a people-centered and process-based framework. Also in representation of Techo, two practitioners took the lead adopting a humanitarian-architecture approach to the design and building process (see Chapter 6, by Martins, Hobeica, Hobeica, & Colacios, in this volume). In the words of Esther Charlesworth (2014, p. 6), this perspective implies "having a concern for, and wanting to help improve the welfare of, people in need" and "using design skills to assist vulnerable communities," particularly after crises engendered by wars or disasters.

In this chapter, shelter provision is understood as a participatory process involving humanitarian designers, volunteers, and affected families. It was possible to highlight the empowerment actions and identify the areas in which more support would be needed, eventually extracting lessons for future rebuilding processes. In the next section, I present Puerto

Rico's disaster context and the involved challenges, before introducing Techo's recovery initiative. Following the method section, I examine Techo's strategy, highlighting the organizational setting and the operational procedures to standardize and implement its kit-based module. Then I highlight some lessons learned from Techo's experience. These insights may support the implementation of future volunteer-led community-based post-disaster rebuilding processes in Puerto Rico, as well as in other similar locations.

2. Post-disaster challenges in Puerto Rico

2.1 The 2017 hurricanes' aftermath and recovery challenges

In September 2017, Puerto Rico, an unincorporated US territory, was impacted, only two weeks apart, by two of the most forceful hurricanes it has ever witnessed since 1928. The combined effects of the category-5 and category-4 hurricanes Irma and Maria were catastrophic, resulting in almost 3000 deaths and extensive damage to housing and infrastructures (Pasch, Penny, & Berg, 2019; Santos-Burgoa et al., 2018). The entire island went offline for more than two months, without electricity, running water, and communication systems. The level of devastation was extreme and the recovery process slow, still ongoing more than two years after the landfalls.

Puerto Rico's insularity, its distance of more than 1600 km to the continental United States, and its lack of adequate resources made the emergency response difficult, expensive, and time-consuming. The construction costs in Puerto Rico are higher than the average costs in the rest of the United States, as most materials are imported. Since 1920, all imports have been subject to the Jones Act, requiring all transport of goods to Puerto Rico to be done only from US ports with US-built, flagged, and crewed ships (COR3, 2018). During and after emergencies, this condition hinders the entry of international aid and affects the availability and cost of highly sought construction materials.

Since 2006, Puerto Rico has also been undergoing a long economic crisis that has engendered precarious conditions for many households. The median income per capita in 2016 was USD11,688, and about 45% of the population lived below the federal poverty level, the largest when compared to the continental United States (COR3, 2018). Estimates indicate that the poverty level increased by up to 60% after the 2017 hurricanes (IEPR, 2018).

Previous to the 2017 hurricanes, the unemployment rate was as high as 11.7%, and the participation in the labor market was as low as about 40%, hindering economic growth (IEPR, 2018). The persistence of these barriers mostly results from ineffective economic strategies geared toward attracting US capital through incentives, higher labor costs, the expiration of industrial tax exemptions, and the increasing public debt exceeding USD70 billion (IEPR, 2018). The economic contractions and impoverishment have resulted in reductions of the island's overall population, with an increase in the elderly, in people living below poverty levels, and in outward migration (RPRAC, 2018).

The long economic crisis also caused the progressive reduction of government personnel, which weakened the support necessary for the maintenance of infrastructure, leading to ineffective official response to the hurricanes' crisis (IEPR, 2018). This fact contributed to deepening the multidimensional challenges that the Puerto Rican population was already facing. The 2017 hurricane season was very active, and by September the federal resources available were already stretched. Many NGOs stepped up and were able to help during the crisis and reconstruction, for months well after the hurricanes.

2.2 Challenges facing the housing stock

Along with the collapse of lifeline infrastructures, the 2017 hurricanes led to the partial or full destruction of a substantial number of houses in Puerto Rico, as summarized in Fig. 2.1. With an average age of around 50 years, the existing housing stock is highly vulnerable to natural hazards and needs retrofitting (RPRAC, 2018). About half of it is informally built or poorly maintained, outside the construction permitting process or without proper land tenure (COR3, 2018). Moreover, this type of informal construction is usually associated with having reduced structural integrity to withstand storms, hurricanes, or earthquakes.

Most of the Puerto Rican families were then dependent on governmental disaster aid, as few of them had homeowner insurance. About 90% of households applied for immediate relief and housing assistance from the US Federal Emergency Management Agency (FEMA).

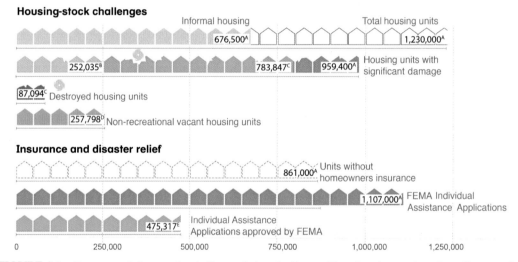

FIGURE 2.1   Summary of data on the challenges facing the Puerto Rican housing stock and on disaster relief assistance. *Credit: Judith Rodríguez Portieles, based on estimates by: (A) COR3 (Central Office of Recovery, Reconstruction, and Resiliency). (2018).* Transformation and innovation in the wake of devastation: An economic and disaster recovery plan for Puerto Rico. *Retrieved from http://www.p3.pr.gov/assets/pr-transformation-innovation-plan-congressional-submission-080818.pdf, (B) FEMA 2018, quoted in RPRAC (Resilient Puerto Rico Advisory Commission). (2018). ReImagina Puerto Rico: Housing sector report. Retrieved from https://reimaginapuertorico.org/wp-content/uploads/2019/05/Housing_Sector_Report_ReImagina_Puerto_Rico_ENG_09.21.2018.pdf, (C) Puerto Rico Government 2017, quoted in RPRAC (Resilient Puerto Rico Advisory Commission). (2018). ReImagina Puerto Rico: Housing sector report. Retrieved from https://reimaginapuertorico.org/wp-content/uploads/2019/05/Housing_Sector_Report_ReImagina_Puerto_Rico_ENG_09.21.2018.pdf, (D) US Census Bureau American Community Survey 2016, quoted in Hinojosa, J., Meléndez, E. (2018). The housing crisis in Puerto Rico and the impact of Hurricane Maria. Center for Puerto Rican Studies report. Retrieved from https://centropr.hunter.cuny.edu/sites/default/files/data_briefs/HousingPuertoRico.pdf, and (E) FEMA (Federal Emergency Management Agency). (2017). Puerto Rico Hurricane Maria (DR-4339). Retrieved from https://www.fema.gov/disaster/4339.*

Yet, only about 40% of those applications have been granted aid (FEMA, 2017). Not having a property title was a barrier, limiting eligibility to receive federal disaster aid. The official land registry records about 70% of property titles, but many are out of date (COR3, 2018).

A year after the disaster, the Puerto Rican Government drafted a recovery plan, dependent on federal disaster aid, with several housing goals. These include retrofitting homes to make them less vulnerable to damage, relocating households from the most dangerous areas, and increasing insurance coverage to help to rebuild after disasters (COR3, 2018). Another challenge identified in the housing stock was the existence of "people without houses and houses without people" (RPRAC, 2018, p. 27). Due to the economic crisis, there is a surplus of non-recreational vacant houses, from which 57% are either foreclosed or abandoned (Hinojosa & Meléndez, 2018). Thus, further evaluations on hurricane damage are needed, as there is a potential for these derelict units to provide housing for displaced hurricane survivors.

2.3 Emergency-housing efforts

In Puerto Rico, there is a lack of preparedness for transitional-housing responses, such as emergency shelters and temporary housing for households who become homeless following disasters or due to economic difficulties (RPRAC, 2018). Municipalities typically use public schools and sports complexes as emergency shelters, but these are often not adequately equipped, and such practices also have side effects (notably the delay of the return to normalcy for school children).

Most of the FEMA (2017) hurricanes disaster-relief aid conceded to homeowners and landlords consisted of temporary blue plastic tarp roofs, expected to last only 30 days. The blue tarps could be installed by the US Army Corps of Engineers or by the families themselves. Many blue tarps continue today to be the roof of many houses. Other provided temporary aid was the Transitional Sheltering Assistance to displaced families staying in hotels located in the continental United States (FEMA, 2020). Long-term assistance for homeowners comprised loans and grants for repairs or replacements.

The hurricanes' aftermath brought a months-long period of adversity that represented more expenses to many families, either working fewer hours or left jobless. In addition to the cleaning and repairing of their houses, families needed to pay for prepared foods, bottled water, and alternative sources of energy (IEPR, 2018). At the refugee center, far from their original communities, it was difficult for the affected families to project into the future and make decisions. Many were living in a situation of fear and instability related to their jobs and belongings.

Techo, a youth-led NGO with more than 20 years of experience in Latin America and the Caribbean, started to work in Puerto Rico in October 2017, targeting families with urgent housing problems (Susmel, 2018). Few weeks after Hurricane Maria's landfall, Techo initiated a short-term project for transitional shelter. The project helped to fill the gap in affected communities in which the governmental disaster response was either inexistent or insufficient. Techo's general work model is based on the creation of a link between volunteers and the targeted community, who join efforts to build solutions for improving the beneficiaries' living conditions. Techo thus seeks to ensure that community members become the transforming actors of their territories (Techo, n.d.).

Techo's transitional-shelter effort was a humanitarian-architecture response that fits into an "all-of-society" approach (IISD, 2018), in which it is critical to share responsibilities among the different stakeholders. It empowered communities and volunteers to collaborate in a fast response that was adapted for the community's recovery, built by them and located in their place of belonging. This type of emergency response helps to develop capacity on aided self-help construction methods, in a process in which potentially anyone in the community can contribute and learn how to assemble a transitional shelter. Techo somehow gave continuity to such a strategy, which had earlier been employed in Puerto Rico, notably in the aftermath of Hurricane San Felipe II, also known as Hurricane Okeechobee in 1928. At the time, the American Red Cross emergency-shelter program distributed free construction materials to approximately 37,000 rural families to rebuild their houses (Crane, 1944).

3. Method

This research adopted the case-study method, which enabled a holistic understanding of Techo's transitional-shelter response in Puerto Rico while retaining its meaningful character-istics (Yin, 2014). Given that the intervention is still recent, the main sources of this study were interviews with Francisco Susmel, architect and project leader in Techo Puerto Rico, and Eliseo Berríos, a local volunteer architect who runs his own firm. The semistructured face-to-face individual interviews were held in August 2019 and included several follow-up sessions.

The insights from the interviews served as a basis for capturing main lessons and were complemented by primary data provided by the interviewees, comprising the following documentation:

— Techo Puerto Rico's unpublished *Construction manual for emergency housing post-Maria* (in Spanish);
— site information and geolocation data of the interventions;
— work-process materials: slideshows, photographs, time-lapse imagery, articles, videos, social-media posts, working notes, and sheets; and
— design materials: digital 3D model and technical architectural drawings (including plans, sections, details, axonometric views, and renders).

As secondary sources, I consulted historical documents and recent governmental reports to understand the context of the hurricanes' aftermath and the socioeconomic conditions in Puerto Rico.

4. Techo's transitional-shelter strategy

Between December 2017 and July 2018, around 600 volunteers joined Techo's activities in Puerto Rico to build 70 timber-framed shelter modules of roughly 36 m^2. According to Sus-mel, Techo built shelters in various locations: the San Isidro and Villa Sin Miedo communities in the municipality of Canóvanas, and other communities in the municipality of Toa Alta. San Isidro and Villa Sin Miedo are impoverished areas mostly inhabited by immigrant families

FIGURE 2.2   A sample of houses destroyed by Hurricane Maria in San Isidro. *Credit: Eliseo Berríos (2017).*

that settled after Hurricane Hugo in 1989. Moreover, Hurricane Maria severely flooded most houses in San Isidro, the focus of this study (Quiñones, 2018). In this community, most families lack property titles and access to infrastructures. Fig. 2.2 shows the local impacts of the hurricane.

4.1 Organizational setting

Techo defined the locations of the interventions according to the immediate housing needs and the communities' levels of poverty and vulnerability. As the main criteria to get involved with and assist a community, Techo considered the number of destroyed houses and the inhabitants' willingness to work with the volunteers. One of Techo's first activities consisted in gathering information in several affected municipalities to reach and select the most poverty-stricken communities. With the help of more than 100 volunteers, Techo conducted family surveys to identify the main housing needs, as well as key communitarian issues, such as health risks, tenure insecurity, and low educational level.

At an early project stage, Techo's team tried to import ready-made units or donated construction materials, but they found many barriers, according to Susmel. Given the scale of the hurricanes' damage and the state of emergency, port operations were halted. Eventually, the high tariffs made it extremely difficult to import any merchandise. The shortage of construction materials at the time was an additional challenge for the realization of this shelter project.

Consequently, Techo conceived transitional-shelter units as minimal and temporary housing spaces to serve the immediate post-disaster necessities. They designed the shelters in such a way as to provide the affected people with a modular and durable housing solution that could later be improved and expanded, or dismantled and reused. The module's internal space can be customized with divisions according to each family's needs, but the shelter would be provided without infrastructural services. Techo expected that eventually the families themselves would upgrade the modules, adding bathrooms, electricity connections, and piped water. In the meantime, the beneficiary families shared kitchens and bathrooms with their extended families or neighbors, or used shared community facilities. The minimal architectural typology helped to speed up the disaster recovery, avoided regulatory barriers, and allowed displaced families to promptly resettle in the territory where they previously resided.

Techo's project was a completely volunteer-led and community-based intervention, as no governmental support was provided. The funds came from individual and corporate donations and from the affected families. To attract volunteers and donations, Techo gave open talks at local universities and posted on social media. To avoid logistical challenges and dispersion in the reconstruction efforts, Techo started building few shelter modules in a phased and concentrated manner, according to the availability of funds and volunteers. Due to the high costs of skilled construction labor in Puerto Rico, the building of the units relied solely on the efforts of volunteers and affected families. Susmel noted that this was a fairly horizontal process, aiming at continuously getting the community involved while relying on labor and talent available among the volunteers.

Building a module took significant efforts (Fig. 2.3). According to Susmel, each shelter module was achieved by a team of six to ten people, in at least six workdays, at a cost of

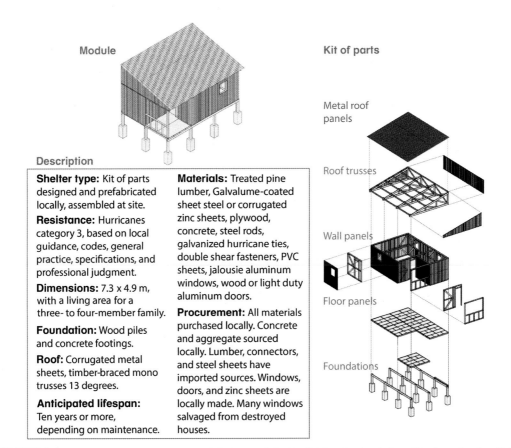

Module

Kit of parts

Metal roof panels

Roof trusses

Wall panels

Floor panels

Foundations

Description

Shelter type: Kit of parts designed and prefabricated locally, assembled at site.

Resistance: Hurricanes category 3, based on local guidance, codes, general practice, specifications, and professional judgment.

Dimensions: 7.3 x 4.9 m, with a living area for a three- to four-member family.

Foundation: Wood piles and concrete footings.

Roof: Corrugated metal sheets, timber-braced mono trusses 13 degrees.

Anticipated lifespan: Ten years or more, depending on maintenance.

Materials: Treated pine lumber, Galvalume-coated sheet steel or corrugated zinc sheets, plywood, concrete, steel rods, galvanized hurricane ties, double shear fasteners, PVC sheets, jalousie aluminum windows, wood or light duty aluminum doors.

Procurement: All materials purchased locally. Concrete and aggregate sourced locally. Lumber, connectors, and steel sheets have imported sources. Windows, doors, and zinc sheets are locally made. Many windows salvaged from destroyed houses.

FIGURE 2.3   Techo Puerto Rico's transitional-shelter module. *Credit: Translated and adapted by the author from Techo Puerto Rico's construction manual.*

around USD8000 per unit. The cost per shelter is about seven to eight times higher than Techo's previous solutions outside Puerto Rico and other international timber-framed shelter products (IFRC, 2011). Yet, according to the interviewees, the lifespan of the Puerto Rican shelter is comparatively longer, expected to be more than ten years, if well maintained. As more volunteers joined and additional donations were received, Techo was able to expand the project.

Unique to Techo's San Isidro team was the incorporation of local architects as volunteers. The design team led the adaptation of the shelter module to the local context, as well as its standardization into a kit of prefabricated parts with assembly instructions. This was achieved through the production of architectural drawings and a construction manual, among others. For Berríos, one of the volunteers, Techo's shelter project became an exercise of leadership and empowerment for architecture professionals. He considered that this project was an opportunity for architects to leave their comfort zone and expose themselves to a different sociocultural environment. Moreover, Berríos claimed that he learned a lot by building with his hands and also gained technical expertise, team-management skills, and knowledge about participatory processes.

4.2 Adaptation of the shelter module

Techo initially brought the shelter prototype used in the 2010 Haiti earthquake response and adjusted it to the Puerto Rican context. The Haitian prototype was already adapted to the Caribbean climatic conditions, including in terms of durability and hurricane resistance (Quiñones, 2018). The design team's modifications involved enlarging the indoor space, positioning partitions according to families' needs, adjusting the module to fit the topography, and taking into account the availability of materials and labor. According to Berríos, due to the Puerto Rican idiosyncrasy, the prototype had to be larger than the Haitian one for it to be a culturally acceptable living space for the beneficiary families. As remarked by Berríos and Susmel, the design of the Puerto Rican shelters needed to achieve the following features:

— a reinforced structure to better resist hurricanes;
— an easy-to-assemble module, with prefabricated parts, which would not require any technical knowledge;
— a replicable module adaptable to different situations, family needs, and sites; and
— economical, fast, and easy-to-produce parts, based on the local availability of materials.

The design and implementation of the modules sought to mitigate as much as possible the hurricane and flood risks. For instance, the shelter design incorporated timber-braced frames with triangulation at the corners, roof trusses, double shear bolts, and hurricane ties for high-wind resistance. The floor structure was raised at least 1 m to prevent flooding, according to the heights informed by the watermarks left by the 2017 disasters in the community.

4.3 Optimization of the construction process

Techo and the volunteers considered the building of the first shelter as a learning experience. Susmel indicated that each module was a joint project between volunteers and families. Volunteers talked with each family individually and invited their members to workshops in which Techo explained its willingness to work with them, the stages of the process, and their responsibilities as beneficiaries. The preparatory stage with each selected family took between six and eight weeks. It included discussion sessions about the best location for the shelter unit and the required adaptations for it to fit the family's needs, and also learning activities about Techo's collaborative construction process. Then Techo engaged more volunteers to support the construction stage, allocated responsibilities, acquired tools, and conducted preparatory workshops on how to perform the job.

One of the key moves of the learning process was the creation of an illustrated construction manual akin to a do-it-yourself booklet. The manual gave an overview of the project's goals, the team roles, and the action plan with families, and broke down the entire construction process into a series of steps. Conceived by the design team, the 24-page manual captured all the optimizations learned from building the first modules and helped volunteers without construction experience to join the project. This tool is an excellent example of how the availability of local architectural talent among the volunteers was capitalized upon by Techo's team.

Yet, the construction manual alone would not be sufficient for the initiative to be a guided participatory process. The manual had to be accompanied by the engagement and training of the inhabitants to build their capacities. Techo fully presented its six-stage shelter-building process in the manual (Fig. 2.4). Each stage consisted of multiple tasks with three to seven steps, which helped to set the roles for collaboration with the community since the project's inception.

As part of the preconstruction stage, volunteers involved the families in three participatory workshops, covering the following topics: role of the family; talents and knowledge; and participatory design. Each workshop utilized graphic templates designed by the team to support engagement and active participation (Fig. 2.5). In the first workshop, the team discussed the different activities that each member of the family could perform and introduced the participants to the collaborative work ahead, connecting their roles to achieve good results as an integrated team. The first template communicated the specific roles and

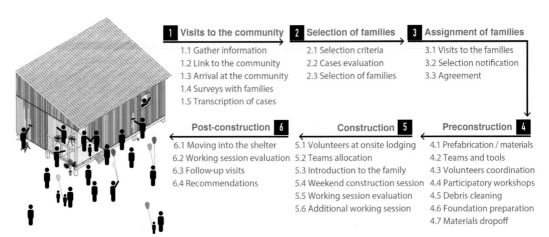

FIGURE 2.4 The construction process of Techo's module streamlined into six stages. *Credit: Translated and adapted by the author from Techo Puerto Rico's construction manual.*

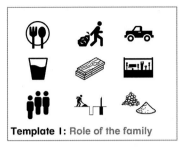  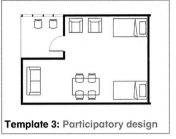

Template 1: Role of the family   **Template 2:** Talents & knowledge   **Template 3:** Participatory design

FIGURE 2.5 Graphic templates used in the preconstruction participatory workshops. *Credit: Translated and adapted by the author from Techo Puerto Rico's construction manual.*

responsibilities that the family would undertake in the preparation of the construction phase. The second workshop helped in the identification of the resources available within each construction team. For instance, each family member, neighbor, and volunteer listed their talents, knowledge, and experiences that could support the building process.

The third workshop, on participatory design, allowed the team to understand the families' spatial needs. Here, the volunteers with an architectural background helped the families to visualize the outcome through several exercises. For instance, the team used a floor-plan template and transparencies to visualize the partitions, windows, and door placement in the space. In addition, the team used other graphic tools to check how the furniture would fit inside the module. Each family could decide about the following to customize their shelter:

— the location of the module at the site;
— the location and size of the balcony (if any), defining the entrance to the module;
— the location of windows, doors, and interior walls;
— the orientation of the roof slope; and
— the height of the raised floor.

The standardization helped the team to streamline the construction of the modules and to embed some flexibility, allowing families to adapt the shelter solution to their needs and lifestyle. The team also helped the families to decide on their shelters' customizations, advising on factors such as ventilation, solar incidence, and waterlogged soils. In addition to optimizing the design, it was important for the team to simplify the construction process into steps to save time and minimize physical efforts. The construction stage constituted an immersive process in which the volunteers lived together within the community during the few days of each building session, usually on weekends. Considering the short timeframe, the team streamlined the process of volunteer work into various tasks, as indicated in Fig. 2.4. In most cases, one weekend was not enough to complete the modules.

To make the construction happen, Techo established clear leadership roles. Work leaders, in charge of organizing and preparing all the construction, planned the logistics and the delivery of construction tools and materials. On the other hand, team leaders monitored the quality of the construction and ensured that the tools and equipment were being safely used. In addition, Techo used prefabricated pieces as much as possible in the module. For the floor, walls, and roof trusses, the team designed timber-framed components that were

FIGURE 2.6   Testing the assembly of the prefabricated parts at Madeco and transporting them to the site. *Credit: Eliseo Berríos (2017, 2018).*

produced by Madeco, a local woodwork vendor. The parts were then transported by truck, stored by the recipient families, and assembled by the team at the site (Fig. 2.6). According to Berríos, few adjustments were still needed in the module to establish the mounting order and the fitting, and to reduce the weight for ease of transportation and assembly.

The construction calendar was divided into four phases, each comprising several tasks presented in the construction manual. The tasks illustrated the related components, average duration, number of volunteers needed, building details, and safety precautions (Fig. 2.7). Phase 1 was for building the foundations, floor beams, and floor panels. This phase was time-consuming and required high-effort tasks, as excavations were necessary for the foundations. Phase 2 was for erecting the wall frames, sills, trusses, purlins, and paneling.

In Phase 3, volunteers and community members installed the roof, designed as a single-sloped surface. The team opted for a mono-pitch roof, instead of a gabled roof (as installed in the first modules), due to the scarcity of anchoring materials in hardware stores and the lack of local expertise in bending steel sheets and making ridges. Berríos indicated that the construction of single-sloped roofs with the prefabricated trusses was easier and safer for the volunteers. Phase 4 concerned the details, such as installing the door and windows assemblies and pouring cement mix on the foundations for safety and waterproofing.

As the families moved into their shelters in the post-construction stage, they often made a celebration with the volunteers. Fig. 2.8 shows two stages of the construction process and the final results already upgraded by the families. Techo included in its work plan an evaluation

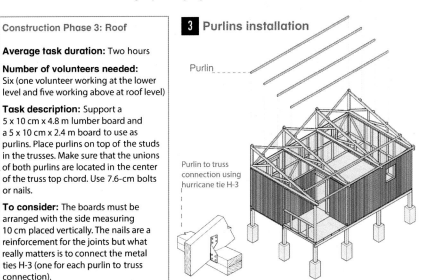

Construction Phase 3: Roof

Average task duration: Two hours

Number of volunteers needed:
Six (one volunteer working at the lower level and five working above at roof level)

Task description: Support a
5 x 10 cm x 4.8 m lumber board and
a 5 x 10 cm x 2.4 m board to use as purlins. Place purlins on top of the studs in the trusses. Make sure that the unions of both purlins are located in the center of the truss top chord. Use 7.6-cm bolts or nails.

To consider: The boards must be arranged with the side measuring 10 cm placed vertically. The nails are a reinforcement for the joints but what really matters is to connect the metal ties H-3 (one for each purlin to truss connection).

3 Purlins installation

Purlin

Purlin to truss connection using hurricane tie H-3

FIGURE 2.7 A sample of the instructions included in the manual. *Credit: Translated and adapted by the author from Techo Puerto Rico's construction manual.*

of the process at the end of the construction. According to the evaluation, the beneficiaries were generally satisfied to be actively included in the process (Quiñones, 2018). The team also made follow-up visits to provide further recommendations to the families regarding upgrading and adding finishes.

5. Lessons learned

This section synthesizes six main lessons that emerged from Techo's experience in San Isidro. When relevant, these insights dialogue with Ian Davis's major lessons from 40 years of shelter experience (Davis, 2011).

Lesson 1: Insularity and island-wide devastation add to the technical and political complexities of recovery processes.

Techo's shelter initiative in San Isidro illustrates the complexities engendered by insularity during disaster recovery. In Puerto Rico, insularity, island-wide devastation, and the lack of preparedness were crucial challenges in the aftermath of the 2017 hurricanes. Remoteness, the unavailability of critical materials, and the disruption of vital social networks in contexts of geographical isolation often hinder efficient recovery operations. In addition, Puerto Rico presents intricate taxation laws and restrictive political ties inherited from its colonial history and its particular administrative status. Being an unincorporated US territory represented an additional hurdle that Techo's project had to face, which nonetheless instigated the team to make the best with locally available resources.

Finding replicable ready-made temporary housing solutions is even more challenging for islands, as the particularities and complexities of these contexts must be taken into account.

FIGURE 2.8   Shelter modules during construction and after upgrades made by the families. *Credit: (A) and (B) by Eliseo Berríos (2017); (C) and (D) by Francisco Susmel (2019).*

Disaster preparedness is crucial for such hazard-prone territories. Ensuring the availability of critical building materials and equipment is a technical challenge shared by islands that needs to be planned ahead. Such a measure can help to avoid the burden of higher prices in the aftermath of disasters. In Techo's project, the engagement of local architects emerged as a valuable resource to navigate these complexities, as they were already familiar with the existing materials' supply chains, logistics, and construction processes.

Lesson 2: Emergency-housing responses should be tied to community-empowerment processes.

Emergency-response processes must be designed in such a way as to help communities to understand their role in rebuilding, by empowering them to make decisions about their future. Through its people-centered approach, Techo's project illustrates a streamlined process of sheltering that is anchored in communities' involvement and common goals toward self-recovery processes (Davis, 2011). The empowerment process was also beneficial for the involved volunteers, given that it represented an opportunity for timely learning experiences and the full exercise of citizenship (Quiñones, 2018; Susmel, 2018).

Lesson 3: Transitional shelter should be part of a more comprehensive risk-informed long-term solution for inclusive and safe housing.

Techo's project was devised as a transitional-shelter response to an emergency, not as a long-term solution. The modules improved the living conditions of the previously displaced families who were ineligible for federal disaster aid and expanded the sheltering offer. However, two years after the hurricane and Techo's intervention, the modules continue to be the homes of many families, as no long-term solution from governmental agencies has come forward yet. Emergency-response planning must take into account the fact that shelters often become permanent housing solutions. Given that Techo's modules can be disassembled or incrementally upgraded by the families, they could easily become permanent homes, as illustrated in Fig. 2.8.

Lesson 4: There is an added value in the inclusion of architects and technical experts in a transitional-shelter response.

The community of housing-related practitioners, in particular architects, urban planners, and engineers, can more actively support rebuilding initiatives led by NGOs, or by public or private actors. The interviewees commented that having local architecture expertise among Techo's team members helped to advance solutions and logistics, as the architects were familiar with suppliers, manufacturers, and costs. In addition, the design team worked on the adaptation of the shelter module, its standardization into prefabricated parts, the definition of construction steps, and the preparation of the workshops.

A noteworthy output of the designers' collaboration was the development of the construction manual, which facilitated the volunteers' work and communicated about the complexities of the participatory process. Nurturing the talent of the volunteers was also important for the realization of the project. These talents were strategically placed in the teams according to people's motivations, to facilitate performing the tasks at hand and avoid exhaustion. The inclusion of local architects helped to build capacities through training and education, as well as social mobilization, as advocated by Davis (2011). It also helped the rebuilding process at the institutional level by maximizing local enterprise opportunities (Sanderson, 2018), as illustrated by the hiring of Madeco to produce the prefabricated parts.

Lesson 5: It is important to balance a project that swiftly fits the community.

Time is essential in a disaster-relief effort, as human lives are at stake. An emergency-housing intervention should aim to achieve a project that has quick positive impacts on the communities. Techo's standardization of the shelter module and the work process supported fast operations and, at the same time, gave enough flexibility for families to adapt it to their needs.

Despite the potential conflict between the post-disaster urgency call to action and the time required to properly achieve a multidimensional local fit, one can argue that Techo's people-centered shelter intervention was aligned with the essential approach of responding to demand rather than supply pressures (Davis, 2011). As stressed by Thompson, "external aid providers need to maintain a humble posture as servant and to furnish their assistance as a vehicle to support the local needs as defined by them, not as a 'deliverable'" (quoted in Davis, 2011, p. 197).

Lesson 6: The best solutions are those that can be immediately implemented with locally available resources.

The best shelter solutions are those that can be implemented with the human and material resources available. It is the best possible response at a given moment. As an example, Susmel mentioned that Techo's team lost time thinking that they could import prefabricated modules.

The team also had to discard more complex architectural design ideas, due to the lack of materials, and as a result, the module is a redone traditional *casa jíbara*, yet it fits well within its context. In addition, they were able to achieve a high-quality construction resistant to hurricanes. In this case, the response had to be quick to protect the displaced families' right to occupy the land in San Isidro, where their support network was located. This is a value that is not measured but cannot be underestimated in an emergency-housing response.

According to Susmel, Techo's project was designed to be open. The housing-module plans are now available for anyone to use and improve in future disaster situations. The shelter modules can also be devised as a system, in which their prefabricated parts are ready to be deployed, or reused after a hurricane or other disaster types. For instance, such optimization exercises provide valuable research and teaching opportunities for universities and NGOs to participate together more effectively and offer support in future disaster-recovery operations.

6. Conclusions

Techo's volunteer-based shelter project in Puerto Rico represents a timely example of how to embed flexibility in a disaster-recovery process and to facilitate aid delivery to the most deprived affected people. These outcomes were possible thanks to the adoption of a people-centered rebuilding approach carried out through the actions of volunteers and the concerned communities. By incorporating design adaptation, standardization, and customization, and by streamlining the construction stages, Techo's project filled a gap in the slow and uneven recovery from the 2017 hurricanes in Puerto Rico.

As indicated in the independent specialized report prepared by the Resilient Puerto Rico Advisory Commission, the 2017 hurricanes have exacerbated an already existing housing crisis (RPRAC, 2018). A transformative reconstruction strategy for safe and healthy housing in Puerto Rico should look for equity in all its dimensions—that is, be pro-poor and inclusive—to move toward a more sustainable and resilient future. Despite the small scale of Techo's transitional-shelter response, it provided a housing alternative for members of some of the most vulnerable Puerto Rican communities who had lost their homes but were ineligible to receive disaster aid. It has also possibly planted a seed for better self-recovery processes in the future.

Acknowledgments

The author thanks Francisco Susmel and Eliseo Berríos for their inputs on the rebuilding efforts, and their openness and availability for interviews.

References

Charlesworth, E. (2014). *Humanitarian architecture: 15 stories of architects working after disaster*. Oxford: Routledge.
Comerio, M. C. (2014). Disaster recovery and community renewal: Housing approaches. *Cityscape, 16*(2), 51–68.
COR3 (Central Office of Recovery, Reconstruction, and Resiliency). (2018). *Transformation and innovation in the wake of devastation: An economic and disaster recovery plan for Puerto Rico*. Retrieved from http://www.p3.pr.gov/assets/pr-transformation-innovation-plan-congressional-submission-080818.pdf.

Crane, J. (1944). Workers' housing in Puerto Rico. *International Labour Review, 49*(6), 608–629.

Davis, I. (2011). What have we learned from 40 years' experience of disaster shelter? *Environmental Hazards, 10*(3–4), 193–212. https://doi.org/10.1080/17477891.2011.597499.

Félix, D., Branco, J. M., & Feio, A. (2013). Temporary housing after disasters: A state of the art survey. *Habitat International, 40,* 136–141. https://doi.org/10.1016/j.habitatint.2013.03.006.

FEMA (Federal Emergency Management Agency). (2017). *Puerto Rico Hurricane Maria (DR-4339).* Retrieved from https://www.fema.gov/disaster/4339.

FEMA (Federal Emergency Management Agency). (2020). *Hurricane Maria. Federal recovery updates.* Retrieved from https://www.fema.gov/hurricane-maria.

Hinojosa, J., & Meléndez, E. (2018). *The housing crisis in Puerto Rico and the impact of Hurricane Maria.* Center for Puerto Rican Studies report. Retrieved from https://centropr.hunter.cuny.edu/sites/default/files/data_briefs/HousingPuertoRico.pdf.

IEPR (Instituto de Estadísticas de Puerto Rico). (2018). *Informe sobre desarrollo humano Puerto Rico 2016* [*Puerto Rico human-development report 2016*]. Retrieved from https://estadisticas.pr/files/Publicaciones/INFORME_DESARROLLO_HUMANO_PUERTO_RICO_1.pdf (in Spanish).

IFRC (International Federation of Red Cross and Red Crescent Societies). (2011). *Transitional shelters: Eight designs.* Retrieved from https://www.sheltercluster.org/resources/documents/transitional-shelters-eight-designs.

IISD (International Institute for Sustainable Development). (2018). *Sendai stakeholders group promotes 'all-of-society' approach to advance Sendai Framework, SDGs.* Retrieved from https://sdg.iisd.org/news/sendai-stakeholders-group-promotes-all-of-society-approach-to-advance-sendai-framework-sdgs.

Pasch, R. J., Penny, A. B., & Berg, R. (2019). *Hurricane Maria (AL152017).* National Hurricane Center tropical cyclone report. Retrieved from https://www.nhc.noaa.gov/data/tcr/AL152017_Maria.pdf.

Quiñones, G. H. (Director) (August 27, 2018). *San Isidro: A flood of hope.* [Video file]. AIA film challenge 2018. Retrieved from https://youtu.be/Cm3qzffw2MA.

RPRAC (Resilient Puerto Rico Advisory Commission). (2018). *ReImagina Puerto Rico: Housing sector report.* Retrieved from https://reimaginapuertorico.org/wp-content/uploads/2019/05/Housing_Sector_Report_ReImagina_Puerto_Rico_ENG_09.21.2018.pdf.

Sanderson, D. (2018). Beyond 'the better shed'. In D. Sanderson, & A. Sharma (Eds.), *The state of humanitarian shelter and settlements 2018. Beyond the better shed: Prioritizing people* (pp. 2–7). Geneva: IFRC/UNHCR.

Santos-Burgoa, C., Sandberg, J., Suárez, E., Goldman-Hawes, A., Zeger, S., Garcia-Meza, A., … Goldman, L. (2018). Differential and persistent risk of excess mortality from hurricane Maria in Puerto Rico: A time-series analysis. *The Lancet Planetary Health, 2*(11), E478–E488. https://doi.org/10.1016/S2542-5196(18)30209-2.

Susmel, F. (2018). *La experiencia del trabajo de TECHO en Puerto Rico* [[*Techo's work experience in Puerto Rico*]]. Retrieved from https://pacosusmel.wordpress.com/2018/03/21/la-experiencia-del-trabajo-de-techo-en-puerto-rico (in Spanish).

Techo. (n.d.). *Modelo de trabajo.* [Work model]. Retrieved from https://www.techo.org/modelo-de-trabajo (in Spanish).

Yin, R. K. (2014). *Case study research: Design and methods* (5th ed.). Los Angeles, CA: Sage.

3

The story of the disaster-relief houses in Iceland

Kristjana Adalgeirsdottir

Aalto University, Helsinki, Finland

OUTLINE

1. Introduction

Forced displacement after disasters, including the adverse effects of climate change, is one of the biggest humanitarian challenges faced by states and the international community in the 21st century. Every year, millions of people are displaced by disasters caused by natural hazards (The Nansen Initiative, 2015). According to a report by the Internal Displacement Monitoring Centre (IDMC), there were seven million new displacements associated with more than 950 disaster events in 102 countries between January and June 2019 (IDMC, 2019). The global community is facing a challenge of immense proportions within all three phases of humanitarian response: emergency, transition, and reconstruction. The process of responding to post-disaster housing needs remains fraught with complexities. As affirmed by Ashdown (2011, p. 25), "providing adequate shelter is one of the most intractable problems in international humanitarian response."

This chapter focuses on a recovery process following a disaster, as opposed to tackling the complexities of shelter response following conflicts. More and more people affected by disasters are forced to live for longer periods in transitional environments: temporary constructions that eventually turn into permanent ones. As an example of the current post-disaster situation, the 2010 earthquake in Haiti left more than one million people homeless, and as of May 2018, nearly 38,000 people (70% of whom are women and children) still live in the same post-earthquake displacement camps (Human Rights Watch, 2019). Graham Saunders (2015), the former head of the Shelter and Settlements Unit of the International Federation of Red Cross and Red Crescent Societies (IFRC), stated that transitional shelters, originally planned as a short-term remedy while longer-term solutions are arranged, have unfortunately often become synonymous with poorly designed and constructed dwellings. Moreover, the transitional-shelter label has been widely utilized to justify such poor solutions and the reality is that they often ultimately become permanent. He continued: "The question that *should* be asked is not whether temporary housing hinders the recovery process, it is why support for recovery and reconstruction is not the default response instead of the provision of temporary housing" (Saunders, 2015).

In this regard, it is interesting to study the housing process that took place following a volcanic eruption on the small island of Heimaey, located in the south of Iceland, nearly 50 years ago. The houses, initially planned during an acute situation as a transitional solution that could eventually serve as permanent houses, have ever since been occupied and adapted to their inhabitants' diverse needs. These houses—known as *Vidlagasjodshus* (meaning disaster-relief houses in Icelandic)—and their story provide a unique opportunity to analyze the factors that proved crucial for their local adaptation during the three phases of disaster response. Thus, this chapter aims to examine the primary factors that have enabled these houses to be still in use and to be considered homes by their inhabitants, five decades after the disaster.

The *Vidlagasjodshus* convey a story of swift responses to unexpected circumstances, an entire community suddenly being forced to flee their homes, alongside how people and houses were able to adapt to the situation. This chapter claims that there are generic factors that apply globally to post-disaster responses that are crucial for culturally integrated recovery. These factors include: the involvement of local stakeholders in the decision-making process; long-term planning from the earliest stages; adjustable housing design with technical, climatic, livelihood, and cultural considerations; an owner-driven process; and a holistic approach to the integrated recovery of the livelihoods and homes of the affected population.

Such research on the *Vidlagasjodshus* is the first of its kind in Iceland. To the best of my knowledge, no academic studies have yet been conducted on the disaster-response process from the building and development perspective, on the progress of the response, or on the cultural adaptation of the imported element houses into homes. The study was conducted with a multiphased, mixed-method approach between 2016 and 2018, utilizing research techniques from both the built-environment and social-science fields. A framework was developed to measure inhabitants' satisfaction with their houses as well as to track the cultural and technical modifications they made to their houses over time. A web-based survey received responses from 433 current and former inhabitants, while six semistructured interviews were conducted and various secondary data were analyzed. This chapter first looks at homes and housing in the context of disaster recovery. It subsequently presents the case study of the disaster-relief houses in Iceland, and then the adopted method. Finally, the results section discusses some generic factors that supported a long-term housing recovery approach in this disaster response.

2. Home, housing, and shelter in disaster recovery

There is an important distinction between the terms *shelter* and *housing* in the relevant literature. Traditionally, shelter is intended to be temporary, even though in practice structures and communities often remain in place far longer than planned. Housing reconstruction refers to rebuilding or providing a permanent solution. A gap exists between emergency relief and longer-term recovery. For surviving households, the sheltering process from immediate protection to permanent housing is a continuous one. However, for the supporting agencies, this process is usually fragmented into discrete phases (relief, recovery, and reconstruction), ultimately undermining longer-term recovery (Davis & Parrack, 2018).

In addition, there is a distinction between housing, a house, and a home. Housing is essential to the well-being and development of most societies. It is a complex asset, with links to livelihoods, health, education, security, and social and family stability. Housing acts as a social center for family and friends, constituting a source of pride and cultural identity (Barakat, 2003).

Throughout the related literature, home is described in various ways, often conflated with or related to a house, a family, and a location. According to the philosopher Kuang-Ming Wu, "home is the intersubjective relationship that brings a self, person, or I into being or existence" (quote in Mallett, 2004, p. 83). This conceptualization of home can, therefore, be understood as fundamental to being. Similarly, the architect Juhani Pallasmaa (1994) stated that from the viewpoints of architectural philosophy and phenomenology, home is not an object or a building but a diffuse and complex condition that integrates memories and images, the past and the present. Home contains the dimension of time, represents a continuum, and evolves as a gradual product of the individual's adaptation and progression through the world.

Studying the homemaking of Georgia's internally displaced Abkhazian inhabitants, Cathrine Brun (2015, p. 44) noted that a "house—the material structure built for human habitation—is not automatically a home. Houses may be turned into homes by their residents, but some houses will never feel like home—never become home." Here, Brun referred to the complex interplay of the occupants' experiences, memories, and relationships with their surroundings as well as the materials, structure, and architecture, which all play a vital role in enabling a house to become a home. The renowned Finnish architect Alvar Aalto also wrote in 1941 about the need for shelters to evolve into permanent homes:

> To satisfy the need for human shelter in an organic way, we must first of all devise a shelter, which will provide the essentials of protection for the individual family and for the community. At the same time, it should be possible for this shelter to develop, step by step, with the social group [...]. In the present situation, there is an

immediate need for an elementary human shelter that can be produced in large quantities. However, at the same time, the permanent character of human life requires that such shelters should be of a nature that they may be developed into shelters on a higher level. That is to say, turned into homes. *(Aalto, 1941, p. 24)*

The notion of "owner-driven reconstruction" has much in common with what some authors call an "aided self-help approach" (Barakat, 2003). Owner-driven reconstruction may be considered the most natural, empowering, and dignified approach toward reconstruction. It encourages people to do what they normally do: build their own homes (Barenstein, 2011). According to Hamdi (2010), donor-driven reconstruction supports instant housing "solutions" that are notoriously inappropriate in layout and technologies, particularly in relation to habits and lifestyles. He argued that participatory processes both get things done in the immediate phase of reconstruction and build capital over the longer term, while delivering more sustainable solutions (Hamdi, 2010). According to findings from case studies on reconstruction projects in Asia and Latin America (Schilderman & Parker, 2014), the beneficiaries' participations in design program, implementation, and monitoring created a sense of community ownership, encouraged housing improvements, and led to the replication of safer techniques (Davis & Parrack, 2018).

The traditional disaster-response approach aims to deliver a finished "shelter product," which is limited in scope and is ultimately inadequate. This approach relies on technical solutions that mainly focus on standardization and speed, with origins in a perception that post-disaster reconstruction is a continuation of the delivery of emergency aid, as if the planning and building of homes are commensurate with the distribution of blankets and other items (Skotte, 2005). The importance of applying a sustainable, holistic approach that recognizes people's housing culture and self-recovery strength is crucial for successful recovery reconstruction. Indeed, as stated by Jennifer Barenstein (2011, p. 194) regarding reconstruction and recovery: "Unfortunately, the consequences of ignoring people's housing culture and livelihoods within the framework of post-disaster reconstruction are coming to light in failed projects all over the world—in abandoned villages, ecological damage, new health problems and dangerous buildings."

In his book *Shelter after disaster*, Ian Davis (1978, p. 33) argued that "shelter must be considered as a process," a series of actions taken to fulfill certain needs, rather than as a mere object, such as a tent or a building. Even though a shift has occurred in recent years toward more process-based approaches in the post-disaster recovery field, this development has been slow. Instead, transitional shelters are often perceived to be "part of nonfood items distribution rather than an ongoing exercise in supporting livelihood, health and security needs" (Davis & Alexander, 2016, p. 183). Discussing the essence of effective shelter and housing following a disaster, Davis (2015) claimed that housing recovery must incorporate a holistic solution to a broad range of challenges, including but not limited to 12 key issues, whereby the absence of only one would reduce the effectiveness of the entire project. These include the affected population's involvement and acceptance; an effective organizational structure within government; the long-term commitment of all supporting groups in both the short and the long terms; technical competence in all key fields including architecture, planning, engineering, and contract management; shelter designs that are adaptable, climatically and culturally appropriate, and environmentally sustainable; and the aim of strengthening the community's economic base (Davis, 2015).

The positive effects of rapidly orchestrated reconstruction toward the long-term recovery of disaster-affected communities have been well documented (Saunders, 2015). However, a separation still persists between the humanitarian sector, which focuses on short-term disaster relief, and the development sector, which works toward long-term recovery and permanent housing development. This gap is also evident in the lack of evidence-based information on how post-disaster reconstruction projects have developed over time in the different recovery phases. Davis and Parrack (2018, pp. 12–13) emphasized the importance of learning from previous projects, revisiting the key results decades after their completion: "Taking a longer-term perspective helps us form a clearer view of recurring themes; […] we still do not know the long-term consequences of different forms of shelter and settlement assistance; this is where we need better evidence from more long-term studies."

3. The Disaster Relief Fund's houses in Iceland

Shortly after midnight on January 23, 1973, a volcano erupted in the Vestmanna Archipelago off the south coast of Iceland. At the time of the event, Heimaey, the largest island of the archipelago, was a fishing village with 5300 inhabitants (2.5% of the Icelandic population) and over 25% of Iceland's annual fishing production (Einarsson, 1974). The eruption forced the entire population to evacuate, instigating the largest migration in Iceland's modern history. Fortunately, it was possible to rapidly evacuate all the island's inhabitants to the mainland. At the initial stages of the eruption, it was impossible to predict its duration, whether the inhabitants would be able to return home soon, or whether the town would even continue to exist. Indeed, as the Icelandic newspaper *Visir* headlined on January 27, 1973: "Will Vestmanna become a modern Pompei?" (Fig. 3.1).

The eruption drew considerable media attention from around the world and relief assistance came from many directions, including the Nordic countries, which offered generous

FIGURE 3.1 "Will Vestmanna become a modern Pompei?" *Credit: Verða Vestmannaeyjar Pompei nútímans? (January 27, 1973).* Visir, p. 1.

financial support. The Icelandic Government established a disaster relief fund (*Viðlagasjóður*) within a few weeks of the eruption, one of the purposes of which was to solve the housing needs of the evacuated islanders. Shortly thereafter, the Disaster Relief Housing Response Committee took actions to assess the possibility of importing houses from the Nordic countries, which could serve as a permanent housing solution if required. Many of the evacuated families had been living with relatives in mainland Iceland or had been assisted in finding rental accommodations, mostly in and around the capital, Reykjavik. Given the continued uncertainty as to the duration of the eruption, many started to look for a permanent place to live in, in the eventuality that return to Heimaey Island would not be possible. As a result, the Disaster Relief Fund organized the introduction of prefabricated wooden houses, which were imported from other Nordic countries. In total, 479 houses were built during the summer and early autumn of 1973 in 20 places around Iceland (Fig. 3.2 and Table 3.1), mostly in the southwest region (Pálsson, 2014). This was an enormous disaster-response task, representing the most extensive housing-construction project ever undertaken in Iceland, accomplished with almost no prior preparation, as underlined by the project leader Gudmundur Thorarinsson during an interview.

On July 3, 1973, the local authorities made a formal announcement that the volcanic eruption had ended. Subsequently, most Vestmanna islanders decided to return home. One year later, approximately half the pre-eruption population had returned to the island and by March 1975 about 80% of the formerly evacuated population had resettled back. As a result, many of the disaster-relief houses were sold on the open market, having served as transitional shelters to the evacuated population. These houses have since been inhabited by different families, each of whom have adapted them to their respective needs and have made them their homes.

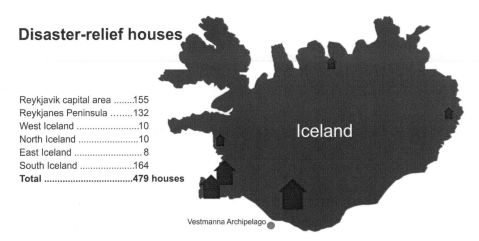

FIGURE 3.2 Distribution of the disaster-relief houses across the different areas of Iceland. *Credit: Based on Pálsson, S. (2012). Eyjamönnum bjargað af götunni: Byggingarævintýri Viðlagasjóðshúsanna [Heimaey islanders saved from the streets: The* Vidlagasjodshouses *building process].* Fréttabréf Verkís, 11, 11–13.

TABLE 3.1 Distribution of the disaster-relief houses by location and donor country.

Area and town	Number of houses	Country of origin
South Iceland	**164**	
Höfn	20	Norway
Hvolsvöllur	5	Norway
Hella	5	Norway
Selfoss	60	Norway
Stokkseyri	12	Norway
Eyrarbakki	12	Norway
Hveragerdi	10	Sweden
Thorlakshofn	40	Sweden
Reykjanes Peninsula	**132**	
Grindavík	42	Sweden
Sandgerdi	10	Norway
Garðar	15	Norway
Keflavík	55	Norway
Hafnafjördur	10	Denmark
Capital area	**155**	
Reykjavik	40	Sweden
Garðabær	30	Finland
Kópavogur	52	Norway
Mosfellssveit	33	Finland
West Iceland	**10**	
Akranes	10	Norway
North Iceland	**10**	
Akureyri	10	Denmark
East Iceland	**8**	
Neskaupsstaður	8	Denmark
Total	**479**	

Credit: Based on Björnsson, B. (1977). Skýrsla um starfsemi Viðlagasjóðs [Report on the activities of the Disaster Relief Fund]. Reykjavik: Viðlagasjóður (in Icelandic).

4. Research method

My fascination with the *Vidlagasjodshus* began many years ago, before I started to study disaster response and architecture. I grew up in Akureyri, a town in North Iceland where ten of the disaster-relief houses were built in 1973. They were located close to my home and I walked by them every day. The houses had a certain mystique. They were in many ways "strange birds": they had arrived from far away and seemed quite different from the other houses in the neighborhood. They were built of wood, painted in dark colors and made of materials and using building technologies that were relatively unknown in Iceland at that time.

Moreover, the story of their inhabitants was captivating. Everyone at the time was aware that the newly arrived villagers had fled their island due to a large-scale disaster, the destruction of their homes leaving them with little more than what they could carry on their boats. They were welcomed and quickly became integrated into the community. The children went to school and the adults to work. Once the eruption had ceased, most of the Vestmanna islanders had the possibility to return to their island, yet the *Vidlagasjodshus* remained. In my annual visits to Iceland over the last decades, I have witnessed the conversion of these houses. They have been transformed from being odd elements in the townscape to houses adapted to the local climate, traditions, and their inhabitants' sociocultural habits. Today, these houses are highly sought-after dwellings, many being situated in prime locations around Iceland. Thus, my research was motivated by the intriguing story behind these houses and that of the disaster response.

Previous studies of the Vestmanna eruption in 1973 focused on the sociological aspects of the community's relocation (Arnardóttir, 2015; Helgadóttir, 2011), as well as on a short-term shelter solution, called "telescope houses," which was tested during the time of the response (Pálsson, 2014). The geological aspects of the volcanic eruption, the progress of the evacuation, and the relief action were covered by various contemporary media and thoroughly documented both in Iceland and abroad. The Icelandic newspapers digitalized all their archives, a valuable resource throughout this study. These materials provided significant information on the scale of the relief, the response tasks, and the reactions of the people involved.

Kortavefur, an Icelandic open-source Internet database (http://www.map.is), archives original building-permit drawings of all the houses built in the country since 1955, as well as all the layout updates, additions, and technical changes that require permits from the local authorities. This database provided an opportunity to explore the evolution and adaptations made to all the *Vidlagasjodshus* from their construction in 1973 to the present day. Analyses of the ensuing drawings led to the selection of four types of houses in different locations, one originating from each of the Nordic countries. The four selected housing types were investigated in detail regarding their interior plans, development, renovations, extensions, and technical upgrading (Fig. 3.3).

I conducted field studies at five locations during the summers of 2016 and 2017. The selected neighborhoods in Reykjavik, Keflavik, Gardabaer, and Kópavogur in South Iceland were visited on one occasion each, and the ten *Vidlagasjods* houses in Akureyri several times during the research period.

An Internet-based survey was created in October 2016 and remained open to collect responses until December 2016. The survey invited all current and former inhabitants to answer

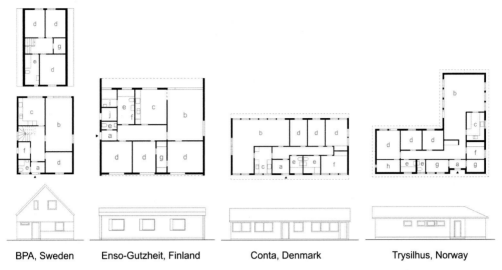

BPA, Sweden   Enso-Gutzheit, Finland   Conta, Denmark   Trysilhus, Norway

FIGURE 3.3 Plans and facades of four house types, indicating the manufacturers and the countries of origin. *Credit: Based on drawings from the database Kortavefur; drawings by Gardar Snæbjörnsson.*

questions about their houses and to share stories and photographs from their lives in them. The questions sought to collect information both on quantitative issues (such as the development of the houses over time, maintenance, costs, and earthquake resilience) and qualitative aspects (the inhabitants' experiences of their houses as homes, issues regarding their qualities, positive and negative experiences, and future plans for the houses). The survey was well received and garnered considerable publicity in the Icelandic media, resulting in 433 participants from all the areas where the houses were built, with an average of 358 answers to each question. The respondents shared many personal memories, including stories of important life events that had taken place in the houses, as well as photographs and links with further information.

The research identified four active Facebook groups connected to the story of the *Vidlagasjodshus*. One comprises owners of the Norwegian houses who share their experiences of their houses' renovations and adaptations. Two groups are composed of childhood friends who lived in the *Vidlagasjodshus* in the 1970 and 1980s. The fourth and largest group (with 2017 members) is linked to an archive project and a webpage (https://www.1973-alliribatana. com), collecting stories, photos, and videos and promoting an active discussion forum.

During the study, I interviewed around 15 house owners in informal discussions while visiting the sites and conducted six semistructured interviews with current and former inhabitants. Given that almost 50 years have passed since the volcanic eruption, it was extremely helpful to interview people who were personally involved in the disaster-relief, response, and reconstruction processes. The leader of the housing operation at the time provided a thorough overview of the process as well as of the political and economic aspects of the response management. Furthermore, a member of one of the construction teams in 1973 offered valuable insights into the technical issues and the necessary adaptation measures associated with the local climatic conditions. One of the directors of the Disaster Relief Housing Response

Committee shared his experiences of the decision-making phases. The other three semistructured interviews were conducted with current inhabitants who have lived in the houses for over 20 years. All the interviews were conducted in 2017. The aim was to understand how the inhabitants had turned their houses into homes and to identify the elements that had been important in this process.

5. Results

The evacuation and housing projects induced by the 1973 volcano eruption represented the largest disaster-response operation in the history of Iceland, accomplished with almost no prior preparation. Moreover, this was the most extensive housing-construction project completed in the country until that time. The impacts of the disaster were widespread and affected every Icelandic citizen to varying degrees.

5.1 The importance of multidisciplinary cooperation

Given the severity of the event, decisions had to be made rapidly, and for the first six-months it was unclear for all the involved parties whether the inhabitants would ever be able to return to their homes. The decision-making process was therefore very challenging regarding the choice between temporary-housing solutions and the permanent relocation of the population. All the interviewed persons emphasized the importance of effective cooperation between the various stakeholders. The coordination was centralized by a local management that knew the culture, circumstances, building traditions, and regulations. This is a fact that was often viewed by both the leader of Disaster Relief Housing Response Committee and the inhabitants as one of the key factors behind the smooth operations.

The discussions about temporary-housing solutions versus permanent houses and the relocation of the whole population were complicated, and the related decision-making process was very complex. The Disaster Relief Housing Response Committee and the local authorities in Heimaey drafted various scenarios and considered both long-term and short-term housing solutions. The decision to import prefabricated wooden houses that could serve as permanent houses was strongly influenced by the importance of perpetuating the evacuees' livelihoods, mostly fishing, in a new location as soon as possible. This fact had to be taken into consideration when the locations of the new houses were selected.

Most of the houses were constructed in areas where the inhabitants could carry on their main sources of income and immerse themselves in the process of recovery. The houses were built in or close to villages that had a similar economic structure and could welcome these new inhabitants. Many communities, especially in the southern coast of Iceland, announced their willingness to support the relocating families. The intention was to strengthen the communities with skilled fishing workers and add to their fishing boat selection in addition to, of course, hosting people in need. The "welcoming attitude" of the host communities indeed played a considerable role in the integration of both people and houses.

In all the phases of the disaster response, representatives of the evacuated population were involved in decision-making. Yet, the uncertainty of the duration of the volcanic eruption and of the possibility of eventually returning home made the coordination of the process complicated. The situation changed often and rapidly as the eruption evolved. The Icelandic saying *"þetta reddast,"* meaning "it will all work out," soon became a mantra for the many overloaded and time-pressured coordinators, engineers, and construction workers faced with this enormous challenge.

Already by the end of February 1973, over 900 out of the 1350 evacuated families had applied to get a donated *Vidlagasjodshus*. Out of the 479 prefabricated wooden houses donated by the Nordic nations, the majority (256 units) came from Norway, with an additional 132 from Sweden, 63 from Finland, and 28 from Denmark (Table 3.1). Two months after the eruption started, the Disaster Relief Housing Response Committee signed the first contract for the import of 200 Norwegian prefabricated houses. The construction of the houses progressed in most cases quickly and smoothly, and the first inhabitants moved in only 100 days after the importation of the houses (Johnsen, 1973) (Fig. 3.4). In many of the small villages where the houses were built, these formed a considerable part of the built environment, for example, in Selfoss (Fig. 3.5).

5.2 Adaptation to local conditions and inhabitants' needs

Climatic conditions and building traditions in Iceland are quite different from the countries where the houses were designed. In modern times, the most common building material in Iceland is concrete. Estimated the material used in over 87% of all houses constructed in the country since 1950. The imported prefabricated wooden houses were built of materials and using techniques that were relatively unknown to Icelanders at that time. Some beneficiaries expressed skepticism to the durability of the houses and how well they would survive in the harsh Icelandic climate. Moreover, given that most outhouses and huts in Iceland are built from wood, a certain stigma existed regarding the use of this material, not deemed "proper" for dwellings. Indeed, one of the comments in the survey stated that *"there was a negative attitude toward these houses. Maybe not officially, but I know that some villagers saw the houses as low-quality wooden huts."*

FIGURE 3.4   Disaster-relief houses under construction in Höfn, South Iceland, in 1973. *Credit: Photography by Vilborg Harðardóttir, with the courtesy of The National Museum of Iceland.*

FIGURE 3.5   Aerial photograph of Selfoss, South Iceland, in 1974, featuring the recently implemented *Vidlagasjodshus* in the front. *Credit: With the courtesy of the photographer Mats Wibe Lund.*

According to the engineers working on the planning and coordination as well as the constructors who built the houses, considerable work was invested into adapting the houses to the Icelandic climate and other local conditions. For example, frequent horizontal rains threatened the wooden walls of the new houses, while the fluctuation of winter temperatures around 0°C, by causing frequent freezing and thawing events, exerted unusual strain on the wooden structures. The engineers paid special attention to insulation, which had to be added to all houses to ensure acceptable indoor temperatures during the winter. In addition, most of the houses were reinforced with steel brackets in the corners to support them against heavy winds and provide for earthquake resistance. The adaptations made during the construction period often proved insufficient, with about 37% of the responses in the survey stating that they added insulation to the houses several years after their completion.

The houses have an area of approximately 125 m^2 (Fig. 3.3). Most of them were one-story, single-family dwellings, except for the houses built in the district of Breidholt in Reykjavik and the row houses in Kopavogur, that have two floors. All the house types had similar, conventional floor plans: a small corridor in the middle of the house leading to the bedrooms, a closed kitchen in one room, one bathroom, and a living room. Thus, they all had clearly defined spaces. During the early 1970s, a period of strong economic growth in Iceland, most of the newly built houses applied similar floor-plan solutions, but they were considerably larger, with an average size of approximately 190 m^2. The *Vidlagasjodshus* were thus relatively small compared to modern Icelandic standards for family houses.

As the house types came from different countries, they carried certain unique traits and traditions from their respective origins. For example, the Norwegian houses featured top-hung, fully reversible windows (called a "housewife's window") that had been developed in Norway. They could be turned 180 degrees, a feature that was totally new to Icelanders. As one interviewee exclaimed: *"Our neighbors really envied me the fine windows, when I turned them outside in and washed from the inside!"* As for the Finnish houses, they came with a sauna, which is an essential part of every Finnish home. In addition, they came fitted with a traditional integrated drying cabinet for dishes above the kitchen sink and a cold storage room for groceries. None of these Nordic inventions became established in Iceland even after being introduced in the *Vidlagasjodshus*. Indeed, Icelanders still wash their windows from the outside and a majority of the saunas have either been integrated with the bathroom or serve as a storage space. As one interviewee noted: *"I think that the sauna has actually never been used, all these 44 years. The previous owners dried their laundry in there, but we have plans to build a Jacuzzi when we expand and renovate the bathroom next year."* Another interviewee stated: *"We had no idea what the intention of that space was. We had never seen a sauna before."*

Studying the development of the houses over time, it is evident that the inhabitants have made major changes to their dwellings (Figs. 3.6 and 3.7). This is also supported by the data from a house study by the planning authorities in Gardabaer, where 12 out of the 16 houses had built extensions and/or major indoor changes that had required building permits from the authorities (Jónsdóttir, 2015). As one of the respondents to the survey stated: *"It was so easy to change the house as needed, because it was made of wood. You could just do it yourself. Take down walls and make additions. And we have done all that."* Many stories were shared about household members armed with electric saws enthusiastically breaking down interior walls and building outdoor decks.

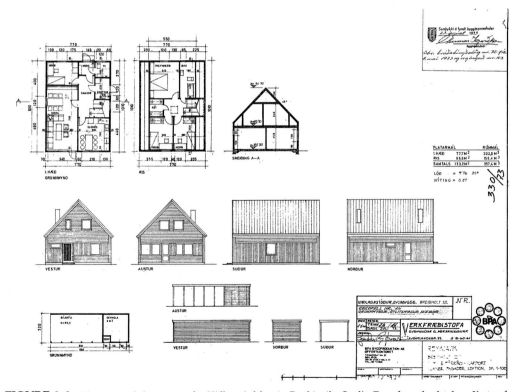

FIGURE 3.6 The original drawings of a *Vidlagasjodshus* in Reykjavik. *Credit: Data from the database Kortavefur.*

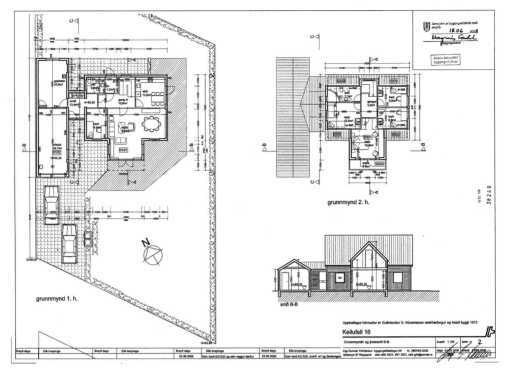

FIGURE 3.7 Drawings of additions and changes of a *Vidlagasjodshus* in Reykjavik. *Credit: Data from the database Kortavefur.*

The repairs and renovations served three main purposes: adapting to local customs, such as integrating the sauna to the bathroom space; reflecting changing family needs over time, such as when bedrooms and kitchens are renovated and bedrooms are connected or divided; and modernizing the houses in line with local standards, such as building winter gardens and sundecks as extensions. The findings from the survey on the various changes and updates made to the houses so far are presented in Table 3.2.

Interestingly, only 3% of the respondents believed that the maintenance costs were higher than those of comparable houses. It was often noted that the houses were made from high-quality wood and that they had endured the Icelandic climate with normal maintenance. When asked about mold, only 26 of the 320 respondents reported having detected any signs of this. More than half of the respondents found the resale value was comparable to houses in the same area. Many mentioned that the *Vidlagasjodshus* were in demand as places to live, especially in established neighborhoods, where they are well maintained.

5.3 Inhabitants' acceptance and satisfaction

It is evident from the survey and the interviews with the inhabitants that they felt at home in their houses. They have managed to transform the foreign standardized houses into

TABLE 3.2 Results from the Internet survey regarding building repairs and development.

Building repairs and development	Quantity	%
Interior development		
Bathroom	211	69.4
Kitchen	206	67.8
Balcony	191	62.8
Garage	141	46.4
Bedroom	139	45.7
Extensions	127	41.8
Living room	97	31.9
Technical adaptions and repairs		
Windows	101	33.2
Water pipes	78	25.7
Roof	72	23.7
Electricity	65	21.4
Isolation	26	8.6
Air condition	10	3.3
No information on repairs available	65	21.4
Total answers	304	n.a.

homes. The group of *Vidlagasjodshus* has become an integrated part of the local neighborhood (Fig. 3.5), and each house manifests the personal style of its inhabitants over time (Figs. 3.8 and 3.9). One of the survey respondents synthesized: "*I grew up in a* Vidlagasjodshus. *My family was evacuated after the volcano eruption. I bought a similar house for my own family when I established one. The houses have a reasonable price for their size, are pleasant to live in and you can easily make the changes you want.*"

The survey responses indicate that the inhabitants viewed the houses positively, with respondents assigning them an average satisfaction rate of 8.6 out of 10. This high rate was also supported in many of the survey's comments and in the interviews with the owners, as expressed in the following quotes: "*The best houses in town. They have their own charm and character that we like, and an interesting story as well!*"; and "*The time we lived in our little house was the best time for our family. Life was simple and our house was a good frame for us and for our children to grow up in.*" Many of the individuals shared stories of important family events related to their houses. The respondents who held a more critical view mostly referred to technical problems, such as heating issues and leakages.

FIGURE 3.8   An example of external adaptations and development of a house in Gardabaer. *Credit: Kristjana Adalgeirsdottir (2017).*

FIGURE 3.9   Another example of external adaptations and development of a house in Gardabaer. *Credit: Kristjana Adalgeirsdottir (2017).*

When asked whether the houses have a negative reputation, 44% answered *"no,"* while only 26% associated some negative conceptions with the houses, largely owed to a lack of knowledge about the qualities and possibilities that the houses offered. As stated by a respondent, *"in the beginning many were skeptical about the houses, [they] were a bit strange, but now they have proved themselves and are regarded as any other house here."* Another person shared: *"My wife did not want to move to a* Vidlagasjodshus, *but now she is the one who does not want to move away."* To the question on whether the inhabitant would like to still be living in the house in five years, 92% answered *"yes."*

6. Conclusions

The findings from the survey and the interviews with inhabitants demonstrate that the houses have passed the test of time and provided their former and current residents with safety and shelter, both as transitional houses in the aftermath of the volcanic eruption as well as permanent homes thereafter.

Similarly, as Cukur, Magnusson, Molander, and Skotte (2005) demonstrated in their studies from Bosnia and as supported by the findings of various other studies (Barakat, 2003; Barenstein, 2011), the experience of the *Vidlagasjodshus* clearly illustrates that understanding the sociocultural aspects of the reconstruction process is one of the key factors behind achieving success. When planning and implementing a reconstruction project, it is necessary to adopt a holistic approach and to understand how the affected people's whole livelihood networks, traditions, customs, and wishes for the future need to be included within the disaster response's physical interventions. The findings of this study of the Icelandic *Vidlagasjodshus* clearly demonstrates, based on the interviews with various stakeholders and in the survey responses, that one of the key factors ensuring the acceptance and adaptation of the imported prefabricated houses was local participation during all stages of the process.

Even though every reconstruction project has different cultural, climatic, and economic conditions depending on its location, the results of this study of the Nordic houses in Iceland provide valid generalizable information and serve as an important contribution to the ongoing discussion. Indeed, the results support previous findings on the factors that are crucial when tackling post-disaster housing problems. First, locally integrated solutions are of utmost importance. Second, the layout design, materials, and technical structures should allow the inhabitants to adjust their houses according to their needs. Third, various local stakeholders should always be involved in the decision-making process. Finally, long-term solutions should be planned from the very beginning of the process.

The story of the *Vidlagasjodshus* demonstrates that, with a clear, unified plan of action and the common intent of all involved parties (including policymakers, aid workers, developers, and inhabitants), it is possible to solve an immediate post-disaster housing challenge. The findings indicate that the provision of simple modular houses can serve their purpose and offer a feasible framework for a home. Moreover, it is possible to find solutions that support a smooth and sustainable transition from the emergency phase to one of a permanent home. After all, houses tell stories—and we must learn to listen to them.

References

Aalto, A. (1941). Research for reconstruction in Finland. *Journal of the Royal Institute of British Architects, 48*(3), 78–83.

Arnardóttir, I. (2015). Það var bara að bjarga hlutum og hamast: Rannsókn efnismenningar á ögurstundu *[One just had to save the stuff and carry on: Study on materialism in emergencies]*. Bachelor's dissertation. Reykjavík, Iceland: University of Iceland. Retrieved from https://skemman.is/handle/1946/21666?locale=en (in Icelandic).

Ashdown, P. (2011). *Humanitarian emergency response review*. Report to the UK Department for International Development. Retrieved from https://assets.publishing.service.gov.uk/government/uploads/system/uploads/attachment_data/file/67579/HERR.pdf.

Barakat, S. (2003). *Housing reconstruction after conflict and disaster*. Humanitarian Practice Network Paper 43. Retrieved from https://www.ifrc.org/PageFiles/95751/B.d.01.Housing Reconstruction After Conflict And Disaste_HPN.pdf.

Barenstein, J. E. (2011). The home as the world: Tamil Nadu. In M. Aquilino (Ed.), *Beyond shelter: Architecture for crisis* (pp. 184–195). London: Thames and Hudson.

Björnsson, B. (1977). Skýrsla um starfsemi Viðlagasjóðs *[Report on the activities of the disaster relief fund]*. Reykjavík: Viðlagasjóur (in Icelandic).

Brun, C. (2015). Home as a critical value: From shelter to home in Georgia. *Refuge: Canada's Journal on Refugees, 31*(1), 43–54. Retrieved from http://refuge.journals.yorku.ca/index.php/refuge/article/view/40141.

Cukur, M., Magnusson, K., Molander, J., & Skotte, H. (2005). *Returning home: An evaluation of Sida's integrated areas programmes in Bosnia and Herzegovina*. Stockholm: Sida.

Davis, I. (1978). *Shelter after disaster*. Oxford: Oxford Polytechnic Press.

Davis, I. (2015). *Shelter after disaster* (2nd ed.). Geneva: IFRC.

Davis, I., & Alexander, D. (2016). *Recovery from disaster*. London: Routledge.

Davis, I., & Parrack, C. (2018). Taking the long view. In D. Sanderson, & A. Sharma (Eds.), *The state of humanitarian shelter and settlements 2018. Beyond the better shed: Prioritizing people* (pp. 9–14). Geneva: IFRC/UNHCR.

Einarsson, T. (1974). *Gosið á Heimaey: Í máli og myndum [The Heimaey eruption: In words and pictures]. Reykjavík: Heimskringla* (in Icelandic).

Hamdi, N. (2010). Foreword. In M. Lyons, T. Schilderman, & C. Boano (Eds.), *Building back better: Delivering people-centered housing reconstruction at scale* (pp. ix–xi). Warwickshire: Practical Action.

Helgadóttir, G. (2011). Fjölskyldur á flótta: Áhrif eldgossins á Heimaey 1973 á íbúa hennar *[Families on escape: The effect of the 1973 volcano eruption in Heimaey on its inhabitants]*. Bachelor's dissertation. Akureyri, Iceland: University of Akureyri https://skemman.is/handle/1946/9521?locale=en (in Icelandic).

Human Rights Watch. (2019). *World report 2019*. Retrieved from https://www.hrw.org/sites/default/files/world_report_download/hrw_world_report_2019.pdf.

IDMC (Internal Displacement Monitoring Centre). (2019). *Internal displacement from January to June 2019: Mid-year figures*. Retrieved from http://www.internal-displacement.org/sites/default/files/inline-files/2019-mid-year-figures_for website upload.pdf.

Johnsen, Á. (1973). *Eldar í Heimaey. [Eruption in Heymaey]*. Reykjavik: Almenna Bòkafélagid (in Icelandic).

Jónsdóttir, B. (2015). *Ásbúð og Holtsbúð Húsakönnun [Study of the houses in Ásbúd and Holtsbúd streets]. Gardabaer: Baejarskipulag* (in Icelandic).

Mallett, S. (2004). Understanding home: A critical review of the literature. *The Sociological Review, 52*(1), 62–89. https://doi.org/10.1111/j.1467-954X.2004.00442.x.

Pallasmaa, J. (1994). Identity, intimacy and domicile: Notes on the phenomenology of home. *Arkkitehti, Finnish Architectural Review, 1*, 14–25.

Pálsson, S. (2012). Eyjamönnum bjargað af götunni: Byggingarævintýri Viðlagasjóðs Viðlagasjóðshúsanna [Heimaey islanders saved from the streets: The Vidlagasjodshouses building process]. *Fréttabréf Verkís, 11*, 11–13. Retrieved from http://www.verkis.is/media/frettabref/Gangverk-1-tbl-2012.pdf (in Icelandic).

Pálsson, K. (2014). *Hús-naedi og Hús-gaedi eftir gos: Lífsgaedin í teleskópahúsunum [Housing and housing qualities after the eruption: Living standards in the telescope houses]*. Unpublished Bachelor's dissertation. Reykjavík, Iceland: Iceland Academy of the Arts (in Icelandic).

Saunders, G. (October 2, 2015). *The opposition's closing remarks* [Online forum comment, Vulnerability, Resilience, and Post-Disaster Reconstruction International Debates]. Retrieved from https://oddebates.com/first-debate.

Schilderman, T., & Parker, E. (Eds.). (2014). *Still standing: Looking back at reconstruction and disaster risk reduction in housing*. Rugby: Practical Action.

Skotte, H. (2005). Arkitekter i utvikling [Architects in development]. *Byggekunst, 87*(8), 21–29 (in Norwegian).

The Nansen Initiative. (2015). *Agenda for the protection of cross-border displaced persons in the context of disasters and climate change*. Retrieved from https://disasterdisplacement.org/wp-content/uploads/2014/08/EN_Protection_Agenda_Volume_I_-low_res.pdf.

Verða. (January 27, 1973). *Vestmannaeyjar Poompeii nútímans?* (p. 1). Visir.

The influence of technical assistance on the adoption of safer construction practices in Nepal

Eefje Hendriks[1], Benjamin Schep[2], Alexander van Leersum[2]

[1]Department of Architecture, Eindhoven University of Technology, Eindhoven, The Netherlands; [2]BuildtoImpact, Group 5 Consulting Engineers B.V., The Hague, The Netherlands

OUTLINE

Enhancing Disaster Preparedness
https://doi.org/10.1016/B978-0-12-819078-4.00004-6

1. Introduction

Humanitarian agencies often provide technical assistance to reconstruct safer housing after disasters, through means such as construction training, demonstration houses, door-to-door technical advice, community orientations, and educational or informative materials. Although technical assistance is expected to enhance the adoption of hazard-resistant construction techniques, it is unclear what its impact is on the knowledge networks and knowledge-acquisition processes in disaster-affected communities (UN-Habitat & AXA, 2019). Evidence from communication outside the scope of humanitarian housing technical assistance is still limited (Maynard, Parker, & Twigg, 2016; Twigg, 2015; Zerio, Opdyke, & Javernick-Will, 2016), but could help to reflect on the impact of provided technical assistance (Opdyke, Javernick-Will, & Koschmann, 2018; Twigg et al., 2017). An estimated 90% of disaster-affected populations reconstruct without humanitarian technical assistance (Morel, 2018), as humanitarian agencies are incapable of covering the large demand for both emergency- and permanent-housing assistance (Development Initiatives, 2018; UN CERF, 2018).

In hazard-prone areas, poorly constructed housing of low-income communities are disproportionally impacted by natural hazards, accounting for 93% of caused fatalities (GFDRR, 2016; Wallemacq & House, 2018). For such vulnerable communities, it is crucial to acquire, understand, and apply hazard-resistant construction knowledge to enhance their safety (Tran, 2015; UNISDR, 2015). Unfortunately, too often communities still reconstruct similar housing as before (Ashdown, 2011; Parrack, Flinn, & Passey, 2014).

During technical assistance, newly introduced stakeholders often play a dominant role in the coordination and implementation of reconstruction efforts, potentially altering already established construction networks. It is a possible disadvantage for long-term disaster risk reduction that knowledge actors are only temporarily available in the communities (Duncan, Scherer, & Wade-Apicella, 2014; Weichselgartner & Pigeon, 2015), as it takes time to appropriate new knowledge to local construction practices (UNISDR, 2015). How construction networks change due to technical assistance is unknown. There is still insufficient research on which knowledge sources and actors communities rely upon when aiming to rebuild safer houses after disasters (Opdyke et al., 2018), and what levels of understanding they lead to. Pathways to enlarge the role of community-based key stakeholders are underexplored.

This study focused on the impact of technical assistance on knowledge acquisition and its outcomes in low-income disaster-affected communities. Therefore, this chapter aimed to answer the following questions:

— What is the influence of technical assistance on the acquisition and understanding of hazard-resistant construction knowledge by community-based construction actors?
— What pathways could increase the influence of community-based key stakeholders as knowledge sources?

This study compared communities with extensive and with limited technical assistance, following the 2015 earthquakes in Nepal. It focused on the reconstruction networks approximately three years following the earthquakes, which allowed to identify potential differences between the two districts that received different levels of housing technical assistance and to reflect upon long-term knowledge-adoption expectations. Conclusions are drawn based on a combination of data from focus-group discussions, household questionnaires using descriptive analysis and social network analysis. The next section discusses the earlier research on construction-knowledge networks following disasters. Section 3 describes the research setting in Nepal, Section 4 contains the adopted methods, followed by the results presented in six subsections.

2. Construction-knowledge networks following disasters

Disasters often disrupt community networks, which are crucial for fostering adaptation to and learning from these events (Adger, Hughes, Folke, Carpenter, & Rockström, 2005). Trust in knowledge-sharing actors positively influences knowledge acquisition and knowledge sharing within community networks (Norris, Stevens, Pfefferbaum, Wyche, & Pfefferbaum, 2008; Tsai & Ghoshal, 1998). However, noncommunity-based actors, such as governmental and humanitarian organizations, have more issues in gaining trust from community members (Wüstenhagen, Wolsink, & Bürer, 2007), as their authority can be contested, and consensus about their approach is often poor (BBC Media Action, 2017a; Kapucu, Arslan, & Collins, 2010). Whereas humanitarian organizations often have high staff turnover harming trust building, governmental agencies are permanently embedded in the social fabric of communities, and can draw from already established channels of communication (Kapucu, Arslan, & Demiroz, 2010), but typically lack the flexibility required in post-disaster situations. Understanding what positions both governmental and humanitarian actors take as knowledge sources compared to other actors can be used to predict performance and behavior outcomes for individual actors or a group of actors (Borgatti, Everett, & Johnson, 2018).

3. Earthquake-resistant construction practices in Nepal

The application of earthquake-resistant construction guidelines was insufficient before the 2015 earthquakes (Giri, 2013; Oven et al., 2016). Although earthquake-resistant construction techniques were recommended in the Nepalese building code (Build Change, 2015), their application was hardly enforced (Oven et al., 2016). Only 5% of the construction activity was supervised or designed by professional engineers (Chmutina & Rose, 2018). In addition, municipalities lacked capacity to verify the application of recommended earthquake-resistant construction guidelines or construction according to design (Chmutina & Rose, 2018; Giri, 2013). The main decisions were made by verbally employed construction workers, transferring all structural risks to homeowners (Chmutina & Rose, 2018).

3.1 Impact of the earthquake

On April 25, 2015, central Nepal was hit by a destructive earthquake with a magnitude of 7.8 and a multitude of aftershocks including a 7.3 aftershock in May. The earthquakes affected 14 out of the 75 districts in Nepal, killing over 8000 people (National Planning Commission, 2015). The lack of earthquake-resistant construction before the earthquakes caused a loss of 67% of the housing stock (NSET, 2017). Over 490,000 houses were destroyed and another 265,000 were damaged (National Planning Commission, 2015). As the epicenter was located in a mountainous and more remote region of the Gorkha district, most of the damaged or destroyed houses (81%) were in rural areas (Hayes et al., 2017).

3.2 Reconstruction program

Reconstruction efforts were challenged by the scale of the earthquake toll and the remoteness of the scattered settlements in mountainous areas. A largely owner-driven reconstruction approach was therefore embraced. Within this programmatic approach, homeowners were responsible for the reconstruction of their houses, with governmental financial support and different intensities of humanitarian sociotechnical assistance. A simplified organizational structure of the housing reconstruction program is presented in Fig. 4.1.

On a national level, the National Reconstruction Authority (NRA) was established by the government to provide guidance and oversight of the housing-reconstruction program. Refinement and implementation of the national strategy were supported by different governmental bodies, NGOs, INGOs, and civil society. Housing-construction standards were established by the Department of Urban Development and Building Construction, and the provision of housing-reconstruction grants to households fell under the responsibility of the Department of Local Infrastructure Development and Agricultural Roads (World Bank, 2016). Complementary to governmental actions, the Housing Recovery and Reconstruction Platform (HRRP) was tasked with the national coordination of the housing reconstruction efforts by nongovernmental partner organizations, assuring coverage and quality of technical assistance and avoiding duplication (HRRP, 2018a).

Nepal was then organized by provinces, subdivided into districts, and Village Development Committees (VDCs, recently renamed as Gaupalika), which were commonly divided into nine wards. At the district level, governmental and humanitarian agencies worked on coordination and implementation activities. In the 14 most affected districts, a District Technical Office temporarily assigned governmental engineers to inspect the housing-reconstruction process and grant financial support in three tranches (NPR300,000 in total—approximately USD2626—for completely damaged houses), conditional to the compliance with seismic-resistant construction principles (Government of Nepal & NRA, 2016). Some

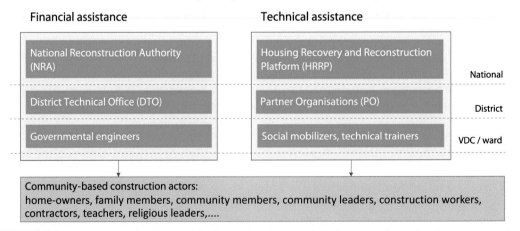

FIGURE 4.1 Organizational structure of the financial and sociotechnical support during housing reconstruction in Nepal. *Credit: Eefje Hendriks.*

of these engineers were hired in collaboration with humanitarian agencies, that supported reconstruction efforts with technical assistance to construct in an earthquake-resistant mode.

Structural decisions in the reconstruction process by homeowners were influenced by family members, other community members, community leaders, construction workers, contractors, teachers, and religious leaders. Contractors made the main structural decisions. Homeowners often participated as construction workers, as shown in Figs. 4.2 and 4.3. Construction workers vary from basically trained masons to untrained day laborers.

3.3 Support tools

For the implementation of the governmental grant program, several technical support tools were developed. A key output was a catalog of earthquake-resistant designs, developed by the Government of Nepal, which included specific guidelines for reconstruction (DUDBC, 2015, 2017), retrofitting (Government of Nepal, 2016c, 2016b, 2016a), and a detailed manual for acceptable exceptions (NRA, 2017). The governmental engineers used an extensive checklist of minimum requirements of technical guidelines to inspect different stages of the reconstruction process (Government of Nepal & NRA, 2016). The checklist included observations of site selection, dimensions, foundation, shape, wall openings, roof, joints, and materials. In case the minimum requirements were not met, governmental engineers used remedial measurements from the correction/exception manual.

Humanitarian partner organizations included a selection of technical-assistance activities, such as community or household orientations, continuous door-to-door technical assistance (mobile technical support), short training for masons, on-the-job training for masons, help-desk or technical resource center, demonstration construction, and community reconstruction committees (HRRP, 2017). Apart from the assistance to the targeted communities, broader awareness was raised with brochures, pamphlets, posters, newsletters, books, hoarding boards, booklets, and training materials (NSET, 2017). Additional awareness was raised via radio and television broadcasting (BBC Media Action, 2017b).

3.4 Variety of reconstruction support across districts

Across the 14 most affected districts, large differences were registered in the types and intensity of the technical assistance. This presumably influenced the sources people drew upon to acquire reconstruction knowledge and likely affected the shape, characteristics, and dynamics of construction-knowledge networks. To gain such insights, an analysis of the type of information and communication, as well as the role of actors in communities, is required. This study compared the most technically assisted district, Gorkha, with the least-assisted district, Okhaldhunga, as shown in Fig. 4.4.

4. Research methods

Wards were selected based on the differing intensity of the received humanitarian technical assistance, as illustrated in Table 4.1. Other selection criteria were as follows:

FIGURE 4.2   Homeowners in Gorkha with their new house under construction. *Credit: Eefje Hendriks (2018).*

considerable earthquake-damage levels, wards with 40 to 250 households, similar socioeconomic demographics, and within reach of a maximum of two hiking days. The primary occupation in both districts was agriculture (75% of the households in Gorkha, 83% in Okhaldhunga), followed by carpentry/masonry (20% in Gorkha, 18% in Okhaldhunga). At the moment of our study, 16 NGOs were active in Gorkha, whereas only 6 in Okhaldhunga.

FIGURE 4.3   Homeowners in Okhaldhunga with their new house under construction. *Credit: Eefje Hendriks (2018).*

The wards were geographically bounded by mountainous landscapes, often separated by several walking hours from other neighboring wards. The sample in Gorkha consisted of eight communities spread over the district, of which three were more remote, high-altitude communities, three middle high-elevated communities, and two more accessible lower communities. In Okhaldhunga, 17 communities spread over the district, also with different degrees of remoteness and altitude, were selected.

The data were collected between February and May 2018, through 1457 household structured interviews, 25 focus-group discussions, and 61 key-stakeholder interviews (see

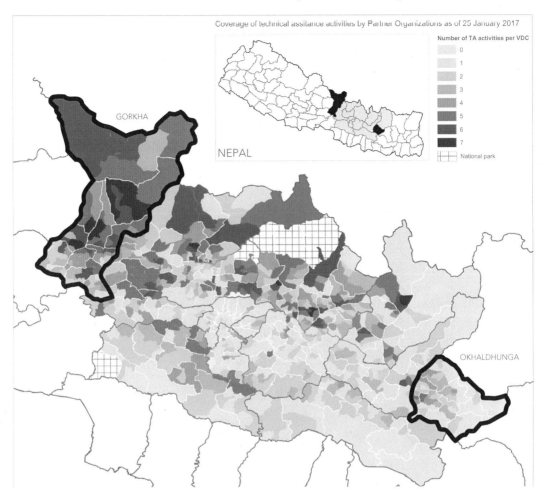

FIGURE 4.4 Coverage of technical-assistance activities in the 12 districts most affected by the 2015 earthquakes in Nepal (data from January 2018). *Credit: Adapted from HRRP. (2018b). Gorkha, interactive dashboard of assistance activities. Retrieved from https://www.hrrpnepal.org/district/gorkha; HRRP. (2018c). Okhaldhunga, interactive dashboard of assistance activities. Retrieved from https://www.hrrpnepal.org/district/okhaldhunga, published with permission.*

Table 4.2). Household interviews and focus-group discussions were mainly used to identify key stakeholders, and to map the use of knowledge sources and measure individual levels of understanding hazard-resistant construction knowledge. Semistructured key-stakeholder interviews aimed at understanding the different roles in the construction-knowledge networks.

Households were selected through stratified random sampling (using wards as strata), assuring a sample size with a 90% confidence level. The household interviews were conducted in each district by three teams of two to three interviewers guided by two to three researchers, during 60 working days in the field. With a duration of 45 to 60 minutes each, these interviews contained only open-ended questions, including one knowledge acquisition (e.g., "*What do you think are reliable sources of information?*"). Each household was asked to identify three actors most significant to support building on an earthquake-resistant mode and specify their relationship to those actors and the use of their knowledge.

TABLE 4.1 Overview of technical-assistance coverage per community.

District	Community	Total number of households	Assistance types	Sum of assistance types	Total technical-assistance interventions
Gorkha	1	77	B, C, J, K, L	5	47
	2	101	B, C, H, K	4	12
	3	224	B, C, F, H, J, K, L	7	60
	4	98	K, L	2	5
	5	250	A, B, C, E, F, G, J, K	8	34
	6	85	B, C, J, K, L	5	46
	7	56	K	1	6
	8	79	C, F, J, K, L	5	36
Okhaldhunga	1	85	A, I, K	3	3
	2	80	K	1	2
	3	61	C, D, K	3	5
	4	113	—	0	0
	5	62	—	0	0
	6	37	K	1	1
	7	76	K	1	1
	8	46	K	1	1
	9	44	A, B, K, L	4	6
	10	42	A, B, K, L	4	5
	11	88	A, K	2	2
	12	95	L	1	2
	13	138	A, K	2	2
	14	48	F, K	2	2
	15	72	—	0	0
	16	80	B, L	2	7
	17	42	—	0	0

Key for assistance categories: *A*, Community reconstruction orientation; *B*, Demonstration construction; *C*, Door-to-door technical assistance; *D*, Government of Nepal housing grant; *E*, Helpdesk/technical support center; *F*, Household reconstruction orientation; *G*, In-kind/material support in housing; *H*, Interactive mobile van/theatre; *I*, Other financial/material/labor assistance; *J*, Reconstruction coordination committee; *K*, Short training; *L*, Vocational training.
Credit: Based on data from HRRP. (2018b). Gorkha, interactive dashboard of assistance activities. Retrieved from https://www.hrrpnepal. org/district/gorkha; HRRP. (2018c). Okhaldhunga, interactive dashboard of assistance activities. Retrieved from https://www.hrrpnepal. org/district/okhaldhunga.

TABLE 4.2 Summary of data and methods.

Research question	Topics	Data			Methods
		1456 household interviews	25 focus groups	61 key-stakeholder interviews	
1	Understanding of hazard-resistant construction knowledge by community members	✔	✔		Descriptive analysis
	Acquisition of hazard-resistant construction knowledge by community members	✔	✔		Descriptive analysis
	Identification and appreciation of key stakeholders	✔	✔		Descriptive analysis, social network analysis
	Knowledge sharing by key stakeholders	–	–	✔	Descriptive analysis, qualitative analysis
2	Self-perceived influence of key stakeholders	–	–	✔	Descriptive analysis, qualitative analysis
	Self-perceived pathways to enlarge role of key stakeholders	–	–	✔	Descriptive analysis, qualitative analysis

Focus-group discussions, separated by gender to assure female representation in the data, aimed to find differences in knowledge access and understanding. Some women are strongly involved in the reconstruction due to the absence of men, who were working in agriculture, urban areas, or even abroad. Participants consisted of a mix of community members willing to contribute to our study. Through interactive mind mapping, the representation of all participants in the focus-group discussions was assured. Names and roles of key actors were collected, and their importance was individually voted for by the participants through dividing ten stones. The general use of knowledge sources was assessed asking a broad open question: *"Where did you get your information from on how to build an earthquake-resistant house?"* Respondents referred to both actors and means of communication and ranked those sources of information based on their importance for the group. The community outcomes were post-categorized, and a weighted district rank was generated. Basic abilities to replicate construction knowledge were tested through a group drawing activity answering *"How do you built an earthquake-resistant house?,"* in which key elements and dimensions were specified.

The interviews and focus groups were the starting point for snowball sampling (Goodman, 1961), identifying at least three key stakeholders in each community for semi-structured interviews. These interviews included questions such as *"Do people apply your technical advice?"* and *"What do you need to reach more people with your advice?"* Answers of the household interviews and focus-group discussions were qualitatively categorized. Frequencies of the arguments were used to compare the two districts using *SPSS*.

Social network analysis (SNA) is often used to analyze social structures (Borgatti, Everett, & Johnson, 2016), through mapping networks of relationships between actors, also called ties, and identification of (hidden) key actors (El-Sheikh & Pryke, 2010; Otte & Rousseau, 2002). In the field of post-disaster recovery, the application of SNA is fairly new (Varda, Forgette, Banks, & Contractor, 2009), but regarded as a unique opportunity to study complex dynamics after a disaster (Kapucu, Arslan, & Collins, 2010). Only a few studies have used SNA in these contexts, for example analyzing the coordination of resources in housing reconstruction following typhoon Haiyan in the Philippines (Opdyke, Lepropre, Javernick-Will, & Koschmann, 2017). Although networks have a dynamic nature, especially in the first phases of post-disaster reconstruction, SNA depicts networks at a specific moment. During our study, the networks were expected to be in a more crystallized stage, and thus less subject to change.

The social structure of the networks was depicted using the number, path, and strength of ties for mathematical and graphical methods (Wasserman & Faust, 1994). Using quantitative data from the household interviews, a knowledge network was built using the SNA tool *Netminer 4*. Metrics were calculated at the actor level, using indegree centrality (IDC), to determine the prominence of actors in the networks (Pryke, 2005). The IDC is the number of connections leading toward the actor, divided by the potential connections within the network (Wasserman & Faust, 1994). More ties imply a more powerful actor (Pryke, 2005). These ties reveal how an actor is embedded in the social structure (Kapucu, 2005), enabling the calculations of some characteristics of the entire network, such as centralization and density. At the network level, using indegree centralization (IDCN), shared ties between actors are visualized showing what knowledge sources households partake during reconstruction. The SNA allowed the visual comparison of networks across the cases (Pryke, 2004, 2005), and facilitated an understanding of the impact and position of actors. Through the IDC, the centrality of the actors was measured, ranking actors in each community and district.

More insights into the role of the key actors were provided using key-stakeholder interviews, which were recorded, transcribed, and translated by Nepalese interpreters and imported into *ATLAS.ti*. Data was systematically coded, characterizing the type of involvement of the participants in the reconstruction, resorting to the most dominant approach for qualitative analysis: grounded theory (Glaser & Strauss, 1999). Two independent coders analyzed 7 out of 61 interview transcripts from both districts, reaching through revision an acceptable agreement score above Krippendorff's Cu $\alpha = 0.8000$. In the key-stakeholder interviews, each actor named their own sources of information, which allowed for the generation of key-actor and actor-category specific data.

5. Research results

5.1 Understanding of hazard-resistant construction knowledge

In both districts, people showed the intention to make changes to their house, which mainly referred to an additional room or floor. It was therefore important to know what level of understanding people had to make those changes in a way that did not hamper the earthquake resistance. In most of the communities, both males and females showed during the focus-group discussions a high ability to remember technical guidelines, to specify them in a drawing, and to point out specific measurements and practical details in the construction steps. The results indicate that in both districts at least an adequate level of understanding to replicate and apply knowledge has been reached for both men and women. The household interviews showed that in the district that received most technical assistance (Gorkha) household members were able to ask significantly more detailed questions about structural safety, and more often considered themselves to be experts on earthquake-resistant construction. People in the district that received limited technical assistance (Okhaldhunga) were less convinced to apply the techniques in the future. In addition, the key-stakeholder interviews indicated that community-based construction actors still have a limited understanding of earthquake-resistant construction techniques for the application in alternative designs, and struggle to analyze or evaluate designs on their earthquake resistance, let alone create new designs. A carpenter in Okhaldhunga stated: *"I need more training on the earthquake-resistant construction of houses. I especially want training on how to build a safe roof, because no one over here can."*

5.2 Acquisition of hazard-resistant construction knowledge

From the different discussions, an overall district ranking of knowledge sources was generated. In addition, household members named again multiple reliable knowledge sources, which were post-categorized, and the frequencies analyzed. Knowledge sources used by community members were not necessarily the same as what was perceived as reliable by household members.

The focus-group discussions revealed different sources used by the communities to obtain information to build in an earthquake-resistant mode, referring to both the means of communication and the advice received from specific groups of actors (see Table 4.3). In Gorkha, community meetings were most important, followed by demonstration houses and technical training. In Okhaldhunga, where technical assistance was limited, people ranked in the first place general information broadcasted by radio and television, followed in the second place by the community leader or local government, and in the third by the community meetings.

Household interviews revealed slightly different perceptions of reliability of the information sources, the most mentioned ones in Gorkha being door-to-door advice and community meetings, followed by written documentation, demo houses, masons, and construction workers. In Okhaldhunga, the engineers' advice was found highly reliable, followed by radio, community meetings, community leaders, and written documentation.

In both districts, community meetings were found to be highly important knowledge sources. This is reflected in the trust in other community members, community-based construction workers, community officials, and family members. In Gorkha, demonstration houses and technical training stood out in importance, and the trust in door-to-door advice by households was high. In the absence of humanitarian technical-assistance activities in Okhaldhunga, the importance of radio and television stood out, followed by community leaders and community meetings. However, the information perceived as the most reliable was the one received from engineers. These results showed the importance of face-to-face communication strategies within the communities and the large reach of knowledge shared via radio and television.

TABLE 4.3   Use and reliability of knowledge sources to build back in an earthquake-resistant mode.

		Gorkha		Okhaldhunga	
		Where did you get information from on how to build back an earthquake-resistant house? (Rank)	What do you think are reliable sources of information? (%)	Where did you get information from on how to build back an earthquake-resistant house? (Rank)	What do you think are reliable sources of information? (%)
Means of communication	Media (radio/television)	5	13.1%	1	35%
	Door-to-door advice	4	46.8%	6	14.1%
	Demo house	2	20%	8	6.9%
	Written documentation (including Internet, pictures)	7	28.5%	4	17.5%
	Community meeting	1	45.6%	3	30.4%
	Technical training	3	0.7%	7	0.5%
Actors	Community leader/local government	6	3.3%	2	21.3%
	Engineer	8	2.6%	5	52%
	Mason/construction worker	–	19.7%	9	9.6%
	INGO/NGO	9	–	–	–

5.3 Identification and appreciation of key stakeholders

This section focuses on the position and role of key actors in reconstruction networks. The categories derived from the different answers given by the participants during the focus groups, presented in Table 4.4. For example, the category "government" was composed of the Government of Nepal, local government, and VDC secretary. Although in Gorkha all communities mentioned engineers as an important actor to build in an earthquake-resistant mode, the most valued sources were community members, followed by local construction workers and humanitarian agencies. In Okhaldhunga, engineers and construction workers were most valued, followed by community members and community leaders.

TABLE 4.4 Identification and appreciation of knowledge sources, based on the answers to the question: *"Where did you get information from on how to build back an earthquake-resistant house?"*

Actors	Identification (% of communities that mentioned this knowledge source)			Appreciation (average % of the votes in each community)		
	Gorkha	Okhaldhunga	Total	Gorkha	Okhaldhunga	Total
Engineers	100	88	94	7.2	24.8	16
Community members	88	75	81.5	27.3	12.2	19.75
Construction workers	63	81	72	15.3	23.3	19.3
INGO/NGO	75	56	65.5	11.2	8.2	9.7
Community leaders	63	50	56.5	4.1	10.7	7.4
Contractor	38	44	41	4.7	5.9	5.3
Government representative	13	50	31.5	3.5	7.9	5.7
Teachers	25	19	22	2	4	3
Health and social workers	25	6	15.5	2.7	1.2	1.95
Material supplier	25	–	12.5	3.1	–	1.55
Loan providers/banks	13	6	9.5	1.1	0.5	0.8
Media	–	19	9.5	–	1.2	0.6
Myself	25	6	15.5	2.2	1.1	1.65
Military	–	6	3	–	0.4	0.2

The results from the SNA were slightly different as they represented quite dissimilar samples and analysis techniques. For each community, a construction-knowledge network was generated identifying the most connected actors, based on the question *"Name a person that is advising you in the reconstruction of your house; specify how you know this person."* Network metrics in both districts clearly showed the strong importance of engineers in post-disaster construction-knowledge networks, with a total mean IDC of 0.432, as shown in Table 4.5. This table also shows the difference in the mean IDC values between the two districts. Contractors, family, community officials, community members, and construction workers were more frequently mentioned as important actors in Gorkha than in Okhaldhunga.

The IDCN shows the distribution of the most dominant actors within the community networks. Clearly, different community networks emerged in the two districts. Some communities had only one single dominant actor, and others had noticeably second- and third-ranking actors. As an example, Fig. 4.5 shows the differences between two communities (one from each of the two districts). At the same time, 9 out of 24 communities had a network with an IDCN of more than 0.500, indicating networks organized around only a few central actors. In Gorkha, 75% of the communities indicated the engineer as the most central knowledge actor. On the other hand, in Okhaldhunga, all communities indicated that the engineer was the most important knowledge actor.

TABLE 4.5   Mean IDC per district.

| Central roles | Mean indegree centrality (IDC) | | |
	Gorkha	Okhaldhunga	Total
Engineer	0.413	0.442	0.432
Contractor	0.265	0.057	0.161
Family	0.225	0.082	0.154
Community official	0.221	0.123	0.147
Community member	0.195	0.111	0.135
Construction workers	0.235	0.095	0.118
Total	0.289	0.209	0.236

Housing technical assistance in the Gorkha district resulted in knowledge networks still organized around the engineers but less dependent on them. In Gorkha, 25% of the communities did not indicate the engineer as the most important actor, whereas the most central role was taken by family members or community members, as illustrated in Table 4.6. The second most important actors were contractors, followed by community officials. The third most important actors were family members, followed by community members and engineers. The second most important actors in Okhaldhunga district were community members, followed by construction workers and family. The third most important actors were community members and community officials (see details of actors' ranking sorted by district in Table 4.6).

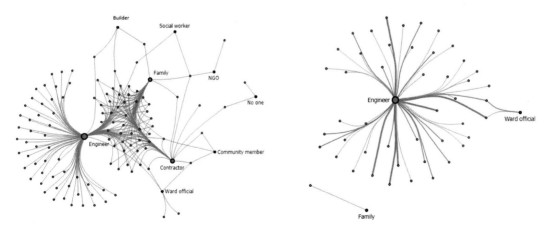

FIGURE 4.5   Example of a construction-knowledge network in two communities: one in Gorkha (left), another one in Okhaldhunga (right). The small gray dots without a denotation represent the interviewed households. The IDCN of the example network in Gorkha is 0.73, compared to 0.82 in Okhaldhunga. *Credit: Benjamin Schep.*

TABLE 4.6 Central actors' ranking sorted by district.

District	Actor	Rank 1	Rank 2	Rank 3
Gorkha	Construction worker	–	12.5%	–
	Community member	12.5%	12.5%	25%
	Contractor	–	37.5%	12.5%
	Engineer	75%	–	25%
	Family	12.5%	12.5%	37.5%
	Ward official	–	25%	–
Okhaldhunga	Construction worker	–	18.8%	12.5%
	Community member	–	37.5%	25%
	Contractor	–	12.5%	18.8%
	Engineer	100%	–	–
	Family	–	18.8%	18.8%
	Ward official	–	12.5%	25%

5.4 Knowledge sharing by key stakeholders

Household members were asked what kind of resources they received from the identified key actors to understand the type of involvement of these in the reconstruction. This section reveals which key actors are most important for the knowledge sharing within the communities, and the reasons behind this, and if they provided additional resources apart from knowledge.

Information was the most frequently shared resource, and originated mainly from engineers, as shown in Fig. 4.6. In Gorkha, about 50% of the people received information from engineers, compared to 70% in Okhaldhunga. Other actors sharing information in Gorkha were family, contractors, community members, community officials, constructors, NGOs, and teachers. In Okhaldhunga, other actors were community members, community officials, family, construction workers, contractors, and teachers. In Gorkha, people primarily depend on family for financial support, whereas in Okhaldhunga financial resources were more spread across community members, family, and engineers. Remarkably NGOs were not found to have an important role in resource sharing, although extensive efforts were invested, especially in Gorkha. Materials were most shared by engineers in both districts, although this was not officially part of their tasks. In illiterate communities, the confusion was common between material suppliers and engineers. The importance of family versus community members differed between the districts regarding material resources.

5.5 Self-perceived influence of key stakeholders

Identified key stakeholders within the communities were interviewed about their role during the reconstruction, which was postcategorized. Overall, the six most prominent groups

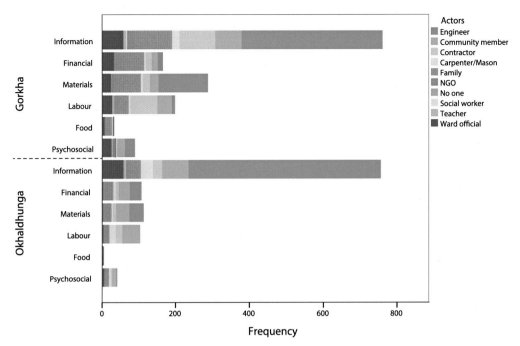

FIGURE 4.6 Overview of resources shared by different key stakeholders in the two districts. *Credit: Benjamin Schep.*

were community officials, government engineers, construction workers, contractors, social workers, and primary- or high-school teachers.

Most key actors suggested that they were considered experts because of their past performance or known experience in the reconstruction (see Table 4.7). Actors were trusted although their skills and knowledge were not yet proven by another earthquake. Governmental actors, such as community officials and engineers, believed that they were considered experts because of their function and official responsibilities, such as providing financial assistance. Other key actors reasoned that their advice was comparable to that of engineers. Most key actors perceived that their advice was applied because they had the trust of households. It was suggested by the interviewees that trust was created by either serving the community or being credible in the community; "*I have done many years of social work in this community, so people trust me and take my advice,*" said a social worker in Okhaldhunga. Government officials in the communities, such as ward leaders, suggested that their advice was applied because of the grant system, forcing households to comply to receive money. However, the advice was also ignored due to limited or too-expensive materials, or because households had already started construction activities before governmental policies were installed.

5.6 Self-perceived pathways to enlarge the role of key stakeholders

The key stakeholders indicated their needs to strengthen their position as a knowledge source. In general, key actors commonly turned to government officials such as a community leader, district engineers or national officials, or people who have received training (see Table 4.8).

TABLE 4.7   Self-perceived influence of key stakeholders.

Statements	Measures and arguments	Mentioned by
Considered expert because of …	Past performance	Engineers, construction workers, social workers, contractors
	Being a trusted servant of the community	Government officials
	Representing the government	Government officials, engineers
	Received training	Construction workers
	Advice is the same as the one of engineers	Government officials, contractors, teachers
Advice applied by others because of …	Grant system	Government officials, engineers, contractors
	Safety following guidelines	Community officials, construction workers, contractors
	Trust	Government officials, construction workers
Advice ignored by others because of …	Limited or too expensive materials	Engineers, teachers
	Construction activities before policies	Government officials, engineers

Contractors and construction workers primarily wanted to enhance their influence through training and interaction with more technically skilled professionals. Government officials expressed the need for better connections to engineers and technically skilled professionals. The knowledge network was significantly top-down oriented with limited consideration of the local knowledge. Engineers were found to learn from other engineers but not from local building traditions or practitioners. Engineers, social workers, and community officials all highlighted that access to government officials was crucial for safer construction practices, allowing local limitations to be taken into account in future programming.

TABLE 4.8   Self-perceived pathways to enlarge the role of key stakeholders.

	Measures/strategies	Requested by
Improve influence through …	More training	Contractors, construction workers
	More training and tools	Construction workers
	Interaction with government officials	Government officials, engineers, social workers, teachers
	Interaction with engineers and technical persons	Government officials, construction workers
	Interaction with trained experienced person	Government officials, construction workers, contractors

6. Discussion

The results demonstrate that while the level of understanding of earthquake-resistant construction techniques was rather high in both districts, households gained a more detailed understanding through housing technical assistance. As construction knowledge was obtained from a wider spectrum of community-based actors in Gorkha, it is more likely that knowledge will be maintained within the communities in the near future, and actors are better able to make changes to their houses. The importance of individual affected households to share knowledge after disasters aligns with findings from earlier SNA research in disaster response (Kim & Hastak, 2018). The role of already trusted engineers could also be used for diffusion of safer construction practices among community-based construction stakeholders. This would be desirable before the governmental support program ends. To define future implications and/or opportunities for the construction knowledge networks, it is also important to identify the second- and third-ranked central stakeholders who are willing to share their knowledge and are already trusted by households. The most important secondary stakeholders in Okhaldhunga are construction professionals, both contractors and construction workers. This study argues for more dialogue between community-based construction stakeholders and governmental engineers to ensure that knowledge is embedded and retained in the construction-knowledge network. Findings of this study highlight the need to train community-based construction professionals to stimulate higher levels of understanding and the effectiveness of radio and television broadcasts and community meetings for knowledge acquisitions in communities.

In Okhaldhunga, without intensive technical assistance, engineers played a dominant role as enabling forces of earthquake-resistant construction, sharing most technical information. Dependence on government engineers is not necessarily a situation that needs to be avoided. However, we perceive that the temporary function of these engineers can overrule the role of construction professionals in the affected areas. Even more, it seems to have reduced their authority. As governmental engineers are only temporarily assigned, their position will change. A wider spectrum of community-based actors is thus crucial as the governmental financial incentive to reconstruct in an earthquake-resistant mode is not permanent (deadlines in this regard were already announced at the moment of this study), and the enforcement of regulations is expected to be reduced. As governmental engineers were only temporarily assigned, their position will change. Humanitarian and governmental organizations should aim to anchor technical knowledge into the construction practices of the local communities and avoid giving such external actors a dominant role in the reconstruction.

The dominance of engineers can be explained by the governmental financial support being conditional in compliance with earthquake-resistant techniques, which limits freedom of choice. Like in the research of Opdyke et al. (2017), government officials had the highest actor centralities. Similar to that study, we found that these roles were the result of obligatory consultations by affected households. Backed by the Sendai Framework (UNISDR, 2015), INGOs prioritize the construction of seismic-resistant housing. The lack of seismic resistance prior to the 2015 earthquakes is supposed to legitimize the role of external stakeholders in the reconstruction. Construction knowledge is used as an instrument for development. In this approach the assumption is made that this technology is of significant added value for households. Decisions are made inside the governmental institutions in Kathmandu. Some households with partly damaged houses were obliged to reconstruct. Limited freedom was given to them to prioritize their development needs.

The government does not actively involve the affected households in the decision-making. Although their ideas were collected and used to develop appropriate design solutions, their influence on policy-making is rather limited. Households are to accept the policies made in Kathmandu. The interviews revealed the desire of households to adapt their (almost finished) dwelling to their own preferences. Unfortunately, before receiving the grants they would not be allowed to do so and were threatened with being placed on a blacklist. This list would exclude their household's next generations from public social services and facilities. The Government of Nepal had already announced the construction deadlines for the three installments at the moment of this study. Some communities expressed severe conflict with the governmental engineers. Households complained about their rigid attitude in following the national regulations and developed mistrust in their expertise. The knowledge use of governmental engineers reproduces the patterns of domination of the Nepalese Government. In Okhaldhunga district, marginalized households decided to take expensive loans to provide for the resources that were not covered by the government. This decision pushed them back in their development, as they could not address their financial resources to meet their most important needs. Remoteness of some communities implied high transportation costs. This aspect increased the price of construction materials, sometimes even to the triple. Prices of construction materials were found to be one of the main barriers. In some cases, community members were already found to ignore the advice of the engineers because of limited availability and high costs of materials, and because reconstruction activities had started even before the government reconstruction grant program was communicated. The decision of the Nepalese Government to provide cash to households to purchase their own materials led to these negative consequences.

7. Conclusion

This chapter took the initial step to explore knowledge networks in post-disaster reconstruction, comparing both technically assisted and unassisted communities in their reconstruction processes. Although collecting extensive data from a disaster case, it is important to validate the knowledge networks and the suggested pathways for their adoption in other contexts. Additional research is needed to identify emergent subgroups of stakeholders who differ in how they communicate technical knowledge.

This research explored construction networks in Nepal, providing insights on the understanding and knowledge acquisition within these networks for the adoption of safer building practices. In addition, this study explored potential pathways to enhance the influence of key actors as knowledge sources. In doing so, the influence of technical assistance on the characteristics of the knowledge networks has been analyzed by comparing two earthquake-affected districts provided with different intensities of technical assistance. Findings of this study recommend humanitarian and governmental agencies to use already effective communication means such as radio, television, and community meetings. The study showed a strong dependence on governmental engineers in the owner-driven reconstruction in Nepal and the need for dialogue with and training of community-based construction actors.

The results show that thanks to humanitarian technical assistance the levels of understanding of earthquake-resistant building techniques have increased. In addition, a higher diversity of stakeholders was found in the construction-knowledge networks. The chapter has underlined the need to critically engage with the role of technical experts. We argue that there is a need for a reevaluation and redrafting of the role of experts and households in housing reconstruction. In a developing country like Nepal, households should have more decision freedom to prioritize their needs in development, in which construction experts can take a facilitative role to support seismic-resistant reconstruction.

Acknowledgments

This independent study was funded through a research grant of the Netherlands Organization for Scientific Research—project number 023.011.055 (*Safer post-disaster self-recovery*). The field research and part of the dissemination have been financially supported by the Catholic Relief Services. Any opinions, findings, conclusions, or recommendations expressed in this study do not necessarily reflect the views of the funding parties.

References

Adger, W. N., Hughes, T. P., Folke, C., Carpenter, S. R., & Rockström, J. (2005). Social-ecological resilience to coastal disasters. *Science, 309*(5737), 1036−1039. https://doi.org/10.1126/SCIENCE.1112122.

Ashdown, P. (2011). *Humanitarian emergency response review*. Retrieved from https://assets.publishing.service.gov.uk/government/uploads/system/uploads/attachment_data/file/67579/HERR.pdf.

BBC Media Action. (2017a). *Nepal's reconstruction communication presentation overview*. Retrieved from https://hrrpnepal.org/uploads/media/NeededCommsforReconstructionNRACommunicationgroup_BBCMediaAction_31stMarch2017_20190506203748.pdf.

BBC Media Action. (2017b). *Nepal Reconstruction Radio Project, overview of aims and elements*. Retrieved from https://www.rethink1000days.org/programme-outputs/khirki-mehendiwali.

Borgatti, S. P., Everett, M. G., & Johnson, J. C. (2016). Analyzing social networks [Review of the book *Analyzing social networks* by S. P. Borgatti, M. G. Everett, & J. C. Johnson]. *Global Journal of Human-Social Science (A), 16*(7), 19−20.

Borgatti, S. P., Everett, M. G., & Johnson, J. C. (2018). *Analyzing social networks* (2nd ed.). London: Sage.

Build Change. (2015). *Build Change post-disaster reconnaissance report*. Retrieved from https://www.buildchange.org/wp-content/uploads/2015/06/2015-Nepal-EQ-Reconnaissance-Report_Build-Change.pdf.

Chmutina, K., & Rose, J. (2018). Building resilience: Knowledge, experience and perceptions among informal construction stakeholders. *International Journal of Disaster Risk Reduction, 28*, 158−164. https://doi.org/10.1016/J.IJDRR.2018.02.039.

Development Initiatives. (2018). *Global humanitarian assistance report 2018*. Retrieved from http://www.globalhumanitarianassistance.org/report/gha-report-2015.

DUDBC. (2015). *Nepal National Building Code. Guidelines for earthquake resistant building construction: Low strength masonry. NBC 203*. Retrieved from https://www.dudbc.gov.np/buildingcode.

DUDBC. (2017). *Design catalogue for reconstruction of earthquake resistant houses. Volume II: Alternative construction materials and technologies*. Retrieved from http://www.dudbc.gov.np/download/cat/13.

Duncan, C., Scherer, S., & Wade-Apicella, S. (2014). *HFA thematic review: Research area 2. Priority for Action 3-Core Indicator 1*. Background paper prepared for the 2015 Global Assessment Report on Disaster Risk Reduction. Retrieved from https://www.unisdr.org/we/inform/publications/49501.

El-Sheikh, A., & Pryke, S. D. (2010). Network gaps and project success. *Construction Management and Economics, 28*(12), 1205−1217. https://doi.org/10.1080/01446193.2010.506643.

GFDRR. (2016). *Building regulation for resilience: Managing risks for safer cities*. Retrieved from https://reliefweb.int/report/world/building-regulation-managing-risks-safer-cities.

Giri, N. (2013). *Implementation of Nepal National Building Code through automated building permit system.* Retrieved from https://www.preventionweb.net/publications/view/35258.

Glaser, B. G., & Strauss, A. L. (1999). *The discovery of grounded theory: Strategies for qualitative research* (1st ed.). New York, NY: Routledge. https://doi.org/10.4324/9780203793206.

Goodman, L. A. (1961). Snowball sampling. *The Annals of Mathematical Statistics, 32*(1), 148–170. https://doi.org/10.1214/aoms/1177705148.

Government of Nepal. (2016a). *Seismic retrofitting guidelines of buildings in Nepal, 2016: Masonry structures.* Retrieved from http://www.dudbc.gov.np/download/cat/12.

Government of Nepal. (2016b). *Seismic retrofitting guidelines of buildings in Nepal, 2016: RCC structures.* Retrieved from http://www.dudbc.gov.np/download/cat/12.

Government of Nepal. (2016c). *Seismic retrofitting guidelines of Buildings in Nepal: Adobe structures 2016.* Retrieved from http://www.dudbc.gov.np/download/cat/12.

Government of Nepal, & NRA. (2016). *Technical inspection guidelines for housing reconstruction.* Retrieved from https://www.preventionweb.net/publications/view/63760.

Hayes, G. P., Meyers, E. K., Dewey, J. W., Briggs, R. W., Earle, P. S., Benz, H. M., … Furlong, K. P. (2017). *Tectonic summaries of magnitude 7 and greater earthquakes from 2000 to 2015.* USGS Open-File Report 2016–1192. Denver, CO: USGS. https://doi.org/10.3133/ofr20161192.

HRRP. (2017). *Core socio-technical assistance package.* Retrieved from https://www.preventionweb.net/publications/view/63753.

HRRP. (2018a). *The path to housing recovery. Nepal earthquake 2015: Housing reconstruction.* Retrieved from https://www.preventionweb.net/publications/view/63580.

HRRP. (2018b). *Gorkha, interactive dashboard of assistance activities.* Retrieved from https://www.hrrpnepal.org/district/gorkha.

HRRP. (2018c). *Okhaldhunga, interactive dashboard of assistance activities.* Retrieved from https://www.hrrpnepal.org/district/okhaldhunga.

Kapucu, N. (2005). Interorganizational coordination in dynamic context: Networks in emergency response management. *Connections: Journal of International Network for Social Network Analysis, 26*(2), 33–48.

Kapucu, N., Arslan, T., & Collins, M. L. (2010). Examining intergovernmental and interorganizational response to catastrophic disasters. *Administration & Society, 42*(2), 222–247. https://doi.org/10.1177/0095399710362517.

Kapucu, N., Arslan, T., & Demiroz, F. (2010). Collaborative emergency management and national emergency management network. *Disaster Prevention and Management, 19*(4), 452–468. https://doi.org/10.1108/09653561011070376.

Kim, J., & Hastak, M. (2018). Social network analysis: Characteristics of online social networks after a disaster. *International Journal of Information Management, 38*(1), 86–96. https://doi.org/10.1016/j.ijinfomgt.2017.08.003.

Maynard, V., Parker, E., & Twigg, J. (2016). *The effectiveness and efficiency of interventions supporting shelter self-recovery following humanitarian crises: An evidence synthesis protocol.* Oxford: Oxfam GB. https://doi.org/10.21201/2016.605179.

Morel, L. M. (2018). *Shelter assistance: Gaps in the evidence.* Retrieved from https://insights.careinternational.org.uk/publications/shelter-assistance-gaps-in-the-evidence?fbclid=IwAR07iegUi_kAqka2un1-0Ji_FnhjLPiYl5q-PIU2ZqQlGYel0yZpSxP_hUas.

National Planning Commission. (2015). *Post disaster needs assessment. Volume A: Key findings.* Retrieved from https://www.preventionweb.net/publications/view/44973.

Norris, F. H., Stevens, S. P., Pfefferbaum, B., Wyche, K. F., & Pfefferbaum, R. L. (2008). Community resilience as a metaphor, theory, set of capacities, and strategy for disaster readiness. *American Journal of Community Psychology, 41*(1–2), 127–150. https://doi.org/10.1007/S10464-007-9156-6.

NRA. (2017). *Correction/exception manual for masonry structure.* Retrieved from https://www.nepalhousingreconstruction.org/sites/nuh/files/2017-06/correctionManual_.pdf.

NSET. (2017). *Safer society. Annual report 2017.* Retrieved from http://www.nset.org.np/nset2012/index.php/publication/publicationdetail/pubfileid-195/pubid-1/type-Recent.

Opdyke, A., Javernick-Will, A., & Koschmann, M. (2018). Household construction knowledge acquisition in post-disaster shelter training. *International Journal of Disaster Risk Reduction, 28*, 131–139. https://doi.org/10.1016/j.ijdrr.2018.02.038.

Opdyke, A., Lepropre, F., Javernick-Will, A., & Koschmann, M. (2017). Inter-organizational resource coordination in post-disaster infrastructure recovery. *Construction Management and Economics, 35*(8–9), 514–530. https://doi.org/10.1080/01446193.2016.1247973.

Otte, E., & Rousseau, R. (2002). Social network analysis: A powerful strategy, also for the information sciences. *Journal of Information Science, 28*(6), 441–453. https://doi.org/10.1177/016555150202800601.

Oven, K., Milledge, D., Densmore, A., Jones, H., Sargeant, S., & Datta, A. (2016). *Earthquake science in DRR policy and practice in Nepal.* Working paper. London: ODI. Retrieved from https://www.odi.org/sites/odi.org.uk/files/resource-documents/10638.pdf.

Parrack, C., Flinn, B., & Passey, M. (2014). Getting the message across for safer self-recovery in post-disaster shelter. *Open House International, 39*(3), 47–58.

Pryke, S. D. (2004). Analysing construction project coalitions: Exploring the application of social network analysis. *Construction Management and Economics, 22*(8), 787–797. https://doi.org/10.1080/0144619042000206533.

Pryke, S. D. (2005). Towards a social network theory of project governance. *Construction Management and Economics, 23*(9), 927–939. https://doi.org/10.1080/01446190500184196.

Tran, T. A. (2015). Post-disaster housing reconstruction as a significant opportunity to building disaster resilience: A case in Vietnam. *Natural Hazards, 79*(1), 61–79. https://doi.org/10.1007/s11069-015-1826-3.

Tsai, W., & Ghoshal, S. (1998). Social capital and value creation: The role of intrafirm networks. *Academy of Management Journal, 41*(4), 464–476. https://doi.org/10.5465/257085.

Twigg, J. (2015). *Disaster risk reduction. Good practice review 9.* London: ODI. Retrieved from https://goodpracticereview.org/wp-content/uploads/2015/10/GPR-9-web-string-1.pdf.

Twigg, J., Lovell, E., Schofield, H., Morel, L. M., Flinn, B., Sargeant, S., … D'Ayala, D. (2017). *Self-recovery from disasters: An interdisciplinary perspective.* Working paper 523. London: ODI. Retrieved from https://www.odi.org/sites/odi.org.uk/files/resource-documents/11870.pdf.

UN CERF. (2018). *2018 underfunded emergencies.* Retrieved from https://cerf.un.org/sites/default/files/resources/cerf_if_overview_20180219_en.pdf.

UN-Habitat, & AXA. (2019). *Supporting safer housing reconstruction after disasters: Planning and implementing technical assistance at scale.* Retrieved from http://urbanresiliencehub.org/housingreconstruction.

UNISDR. (2015). *Sendai Framework for Disaster Risk Reduction 2015–2030.* Retrieved from https://www.preventionweb.net/files/43291_sendaiframeworkfordrren.pdf.

Varda, D. M., Forgette, R., Banks, D., & Contractor, N. (2009). Social network methodology in the study of disasters: Issues and insights prompted by post-Katrina research. *Population Research and Policy Review, 28*(1), 11–29. https://doi.org/10.1007/s11113-008-9110-9.

Wallemacq, P., & House, R. (2018). *Economic losses, poverty and disasters (1998–2017).* Retrieved from https://www.unisdr.org/we/inform/publications/61119.

Wasserman, S., & Faust, K. (1994). *Social network analysis: Methods and applications.* Cambridge, MA: Cambridge University Press. https://doi.org/10.1017/CBO9780511815478.

Weichselgartner, J., & Pigeon, P. (2015). The role of knowledge in disaster risk reduction. *International Journal of Disaster Risk Science, 6*(2), 107–116. https://doi.org/10.1007/s13753-015-0052-7.

World Bank. (2016). *Nepal Rural Housing Reconstruction Program: Program overview and operations manual summary.* Retrieved from http://documents.worldbank.org/curated/en/135481468187745015/Nepal-Rural-housing-reconstruction-program-program-overview-and-operations-manual-summary.

Wüstenhagen, R., Wolsink, M., & Bürer, M. J. (2007). Social acceptance of renewable energy innovation: An introduction to the concept. *Energy Policy, 35*(5), 2683–2691. https://doi.org/10.1016/J.ENPOL.2006.12.001.

Zerio, A., Opdyke, A., & Javernick-Will, A. (2016). Post-disaster reconstruction training effectiveness. In J. Kaminsky, & V. Zerjav (Eds.), *Engineering project organization conference* (p. 16). Washington, DC. Retrieved from https://ses.library.usyd.edu.au/handle/2123/19581.

5

Participatory design for refugee shelters: An experiment in Syrian camps in Jordan

Lara Alshawawreh

Faculty of Engineering, Mutah University, Karak, Jordan

1. Background

Toward the end of 2018, the number of forcibly displaced people worldwide was approximately 70.8 million, exceeding the figure of the previous year by more than 2 million. Out of the 70.8 million, there are about 26 million refugees, 41 million internally displaced people, and more than 3.5 million asylum seekers. However, only about 20.4 million refugees are registered within the mandate of the UN high commissioner for refugees (UNHCR), while the other 5.5 million are the Palestinian refugees registered within the United Nations Relief and Works Agency (UNHCR, 2019b).

The Syrian civil war that erupted in 2011 has caused the influx of the largest number of refugees in the world and was described by the UN High Commissioner for Human Rights as the worst human-driven disaster since World War II (Siegel, 2017). It has resulted in having approximately 5.7 million UNHCR-registered Syrian refugees until November 2019 (UNHCR, 2019f). Most of the refugees fled to neighboring countries including Turkey, Lebanon, Jordan, Iraq, and Egypt. By the end of 2017, the number of Syrians in Jordan was estimated by the Jordanian Ministry of Planning and International Cooperation as 1.3 million (Ministry of Planning and International Cooperation, 2017). However, as of November 2019, only about 654,000 Syrians were registered as refugees. Although about 123,000 of the Syrian refugees in Jordan live in camps, the other 531,000 are living within hosting communities (UNHCR, 2019f). The Syrians in the camps are unauthorized to live in a hosting community without having a Jordanian guarantor. This condition resulted in having Syrian refugees of poor socioeconomic status encamped. Initially, the residents were not allowed to work outside the camps. But by mid-2017, the Jordanian Government allowed the camps' residents to have work permits and access the available jobs in the country (International Labour Organization, 2017). The majority of Syrian refugees in Jordan originated from Daraa city (48%), followed by Homs with 19%, Aleppo 10%, Rural Damascus 9%, and Damascus 8%. Most of them, about 69%, stated that they came from rural areas (Tiltnes, Zhang, & Pedersen, 2019) (Fig. 5.1).

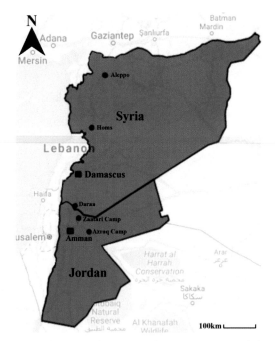

FIGURE 5.1 A map showing the location of Zaatari and Azraq camps, and the main source cities of Syrian refugees in Jordan. *Credit: Lara Alshawawreh (2020), based on Google Maps.*

The temporary condition of the refugee shelters is a requirement that most hosting countries, including Jordan, impose when hosting refugees. The existing standards, such as the *Sphere Project* (Sphere Association, 2018), call for providing shelters that protect the privacy and dignity of their residents. However, the temporary status of the shelters, among other reasons, makes meeting these standards hard to implement. On the other hand, since 85% of the refugees are hosted in developing countries (Devictor & Do, 2016; UNHCR, 2018a), turning the refugees into permanent residents is beyond what these countries can handle.

1.1 Zaatari camp

In July 2012, the need to host large numbers of Syrian refugees in Jordan impelled the UNHCR and the Jordanian Government to set up Zaatari camp. It is located in the Mafraq governorate, about 13 km from the Syrian border in northern Jordan (Fig. 5.1). The camp was set up in nine days on a land that is owned by the Jordanian armed forces (USA for UNHCR, 2017). It involved a rapid deployment of tents as an urgent response to the refugees' influx. In less than ten months, the camp evolved to enclose its current area of approximately 5.3 km^2 (UNHCR, 2018b).

Zaatari camp is divided into 12 districts, and each district has numbered blocks and named streets. Although the universal guidance directs toward having a maximum of 20,000 refugees in any camp setting (UNHCR, 2019a), with a surface area of 45 m^2 per person (Sphere Association, 2018), Zaatari camp reached its peak during April 2013 with more than 200,000 residents, which equals 10 times the recommended maximum number of residents. Moreover, the 200,000 residents occupied less than half of the recommended area. Consequently, in April 2014, the Jordanian Government closed the unofficial border crossings in Daraa (Ledwith, 2014) and opened Azraq camp. Since June 2014, the number of residents in Zaatari camp decreased gradually with time, reaching approximately 75,000 residents by April 2019 (UNHCR, 2019f). This number still exceeds the recommendations for the maximum number of residents per camp but fulfills the minimum surface area per person.

The first sheltering response in Zaatari camp was essentially tents, with only a few prefabricated shelters. However, the tents were unsuitable for the winter months in Jordan as they were prone to flooding (Gatter, 2018; REACH, 2014). The winter seasons of 2013 and 2015 were specifically hard on the residents of the camp, as heavy rainstorms and snowstorms hit the country. Tents were flooded and families were moved to their relatives' shelters, to mosques, to emergency shelters inside Zaatari camp, or were relocated to other camps (Gavlak, 2013; Maayeh, 2015). This is why, toward the end of 2015, most of the tents were replaced with prefabricated shelters (REACH, 2014). Although the prefabricated shelters (known locally as caravans) in Zaatari camp were introduced during the first few months of the camp's opening, the process was only accelerated when the weather storms hit the camp.

The dimensions of the prefabricated shelters vary around the camp due to the diversity of donors. Some shelters were noted to be 4 m × 2.5 m, others were 5 m × 2.5 m. Each shelter hosts up to six family members, and a second shelter was distributed for larger families. However, due to the lack of private facilities, most of the large families (with more than six members) have assigned one of the two given shelters for private facilities (kitchen, shower, and toilet), and the whole family members lived in the second shelter. Nevertheless, all shelters had insufficient spaces that cannot fulfill the needs of the users (Alshawawreh, 2019). The prefabricated shelters are made of 40-mm sandwich panels; the outer skin of the panels is 0.35-mm steel sheets, whereas the inner skin is either steel or timber, and the insulation in between is polyurethane (Albadra, Coley, & Hart, 2018).

ACTED (2017) stated that five years following the opening of Zaatari camp, most of the shelters were in a critical situation. This report showed that the shelters particularly suffered from improper flooring, water leakages, indoor flooding, and rusted walls. ACTED (2017) described the self-built private toilets as sanitary threats as they did not fulfill the minimum hygienic standards. Additionally, Alshawawreh, Smith, and Wood (2017) highlighted how the cultural inadequacy of the shelters was a major concern for the camp's residents. The latter study clarified how the residents have taken many decisions regarding space management that affect their health, but better protect their privacy, such as covering the windows, self-building private toilets, and cooking inside the shelters. Today, eight-years following the establishment of Zaatari camp, most shelters have self-built extensions that are made of corrugated sheets and pieces of canvas. These extensions might temporarily fulfill the needs of the residents, but they do not consider the macro-planning of the camp and therefore could turn it into a future slum. Additionally, there are many facilities in the camp, such as shops, schools, hospitals, mosques, and others, provided by various nongovernmental organizations (NGOs), which have transformed the camp into a city on its own. Moreover, some residents have self-built shops in the camp that became their main sources of income.

1.2 Azraq camp

Azraq camp was opened in April 2014 due to the increasing number of refugees in Jordan. As Fig. 5.1 shows, it is located 20 km from Azraq city, 80 km southeast of Zaatari camp and 90 km from the Syrian borders (CARE International, 2015). The camp was purpose-built and designed to take account of lessons learned from Zaatari camp, such as resorting to T-shelters (a term that is used to describe both temporary and transitional shelters, which usually offer an enhanced political acceptance due to their flexibility (IFRC, 2013)) instead of the prefabricated shelters. The T-shelter has a less complicated structure, which allows the local labors and refugees to participate in the building process. At the same time, it was a cheaper option that saves the transport and production costs that accompany the prefabricated shelters (IFRC, UN-Habitat, & UNHCR, 2014). Moreover, strict regulations against making any amendments to the shelters were established, inspired by the unplanned amendments that were done by the residents of Zaatari camp.

Another lesson learned was in regard to the planning of the camp, as Azraq camp was designed to involve six zones, called villages, each with its own facilities to preserve the community structure that distinguishes the Syrian culture (Gatter, 2018). However, when the camp was initially opened only villages three and six were occupied by the refugees (UNHCR, 2019e). Villages two and five were opened in later stages, while villages one and four were never opened. Each village in the camp is divided into blocks that consist of streets and numbered shelters. Azraq camp was planned to have 13,500 T-shelter units

(IFRC, UN-Habitat, & UNHCR, 2014). It has the potential capacity of accommodating 120,000–130,000 refugees in its total area of 14.7 km^2 (UNHCR, 2019d), which is against the general recommendation of having a maximum of 20,000 residents in any camp settlement (UNHCR, 2019a). According to UNHCR data, the camp reached its peak during July 2014 with approximately 55,000 residents, despite its maximum current capacity of hosting only 50,000 (UNHCR, 2019f). But since June 2018 (and at least until November 2019), the number of residents in Azraq camp has stabilized between 41,000 and 39,000 (UNHCR, 2019f), and the number of used shelters was less than 9000 (UNHCR, 2019d). However, the current number of residents equals twice the maximum recommended in any camp settlement (UNHCR, 2019a).

Similar to the plan of Zaatari camp, the plan of Azraq camp also has a grid system, where rows of white shelters mark up the streets. However, in contrast to Zaatari camp, the NGOs' "base camp" offices in Azraq camp are distant from the shelters. They are located at about ten-minute driving distance to the nearest "village" inside the camp. Gatter (2018) argued that the emptiness of Azraq camp, meaning the unoccupied shelters that separate the villages and the abundance of space, is intentional and aims at limiting the movement of the refugees, to make potential demonstrations and undesired gatherings harder to occur.

The shelter has an area of 24 m^2 (IFRC, UN-Habitat, & UNHCR, 2014) that serves a family of up to six members. The indoor shelter size allocates 4 m^2 per person. This area aligns with the recommended minimum area per person of 3.5 m^2 proposed by the Sphere Project (2011). However, it does not fulfill the needs of the people in the camp as proved by the outdoor spaces that were enclosed by the residents, and by the residents' continuous demands to build extensions. By highlighting this issue, we sustain the questioning of the origin and validity of the 3.5 m^2 standard proposed by the *Sphere Project* (Kennedy & Parrack, 2013). The T-shelter has an interlocking steel structure, covered by 10–15 mm of aluminum-foam insulation, and has external and internal Inverted Box Rib (IBR) metal cladding and flashing. The interior of the shelter includes an additional roofing layer of plastic sheeting. Regarding the floor, a concrete layer was poured over a metal rebar, which transformed the shelter into a permanent structure. Additionally, adjustable footings were used to level up the structure (IFRC, UN-Habitat, & UNHCR, 2014; UNHCR, 2015, 2016).

When the T-shelter was first implemented, the major drawbacks that were highlighted by IFRC, UN-Habitat and UNHCR (2014) included the inability of the IBR to be sealed off and the high amount of heat gain. However, when REACH (2015) conducted an assessment study, the lack of privacy was the main concern of the residents who demanded to have private showers, private outdoor kitchens, outdoor fences, and private outdoor sitting areas. Storage spaces and an additional window were also among the expressed needs.

Although the strict policies in Azraq camp prevented the residents from doing major amendments to their shelters, they succeeded in doing small amendments, such as cutting additional windows, removing the internal roofing and using it to enclose outdoor private areas, and creating a space for a private kitchen (Alshawawreh et al., 2017). Moreover, the residents added many functions to their shelters that were not planned for, such as showering, as the communal showers were considered unsafe and culturally inappropriate (Alshawawreh et al., 2017).

2. Participatory design

Community participation in post-disaster situations has received more attention in recent years, but it has not yet been implemented widely. Earlier projects have had end users involved in the building process or post-sheltering feedback, but rarely during the design phase. Most of the cases that were documented in the recent *Shelter projects 2015—2016* book highlighted the lack of community engagement as a weakness (Global Shelter Cluster, 2017). This gap involved the absence of users' involvement such as in the case of the Gaza project, or the lack of staff training on how to plan community involvement such as in the case of the Tanzania project. Other cases lacked funds, safety, space, or other influencing factors that affected the ability to engage the communities in the shelter design and construction.

Participatory design (PD), also called "cooperative design" and "codesign," has been proposed since the 1970s as a method to fulfill the concept of designing "with the people," not "for the people." Roth (1999) considered PD as one manifestation of the participatory research described as a human-centered design research, to seek better and successful solutions. Carroll (2006, p. 7) defines PD as "the direct inclusion of users within a development team, such that they actively help in setting design goals and planning prototypes."

The evidence of previous studies on refugee shelters using PD as a research method is limited. However, the method was adopted by Architecture Sans Frontières UK who carried out workshops around the world aimed at building communities using PD. Their work in Los Pinos, Ecuador, and in Kenya are two remarkable examples (Frediani, De Carli, Ferrara, & Shinkins, 2013; French, 2011). In the Ecuadorian case, a two-week workshop was undertaken in the municipalities of Quito and Mejìa to explore options for the future upgrading of Los Pinos. The "dwelling" was one of the aspects they explored through different phases: diagnosis, dreaming exercises, and then consolidation of the findings. The dreaming phase included "dreaming through drawing" and "dreaming through modeling." The latter indicated that modeling was a more accessible tool to the residents than drawing. The researchers provided to the participants a kit of several room sizes and asked them to select rooms and build their own dream house (Frediani et al., 2013). The "dwelling" approach in the Kenya workshop was similar to that of Los Pinos. According to French (2011), the four undertaken stages of the experiment were as follows: "diagnosis through walking and talking," "dreaming through drawing," "dreaming through modeling," and "dreaming through typologies."

According to Sanders, Brandt, and Binder (2010), the main challenge with PD is to find suitable techniques that are easy to use by non-designers and allow them to add their unique inputs. These authors organized the PD tools and techniques through a three-dimensional framework containing the form, the context, and the purpose. In addition, Sanders et al. (2010) recommend several variables that could be used to determine the three aforementioned dimensions, such as the group size and composition, the PD to be conducted face to face or online, the chosen venue, and the stakeholder relationship.

This chapter will analyze and discuss a PD experiment held in Jordan, in the Zaatari and Azraq Syrian camps. Following the framework of Sanders et al. (2010), the present study chose the form of 3D mockups through face-to-face working groups to fulfill the purpose of generating design ideas and understanding the refugees' current experience.

3. Methodology

The aim of this research was to understand the space needs and priorities of the Syrian refugees. The adopted methodology was to conduct PD sessions in both Zaatari and Azraq camps. These two camps were chosen due to their location in Jordan, a Middle Eastern country with a long history of receiving refugees. Additionally, Zaatari camp was chosen in view of its large number of residents, which made it the largest refugee camp in the Middle East and one of the largest in the world. The decision of undertaking PD sessions in both Zaatari and Azraq camps aimed at understanding if the experience of people from the same background living in differently designed camps would influence their choices and their perspectives toward their needs and aspirations regarding the spaces they occupy.

Visits to Zaatari and Azraq camps in Jordan have been arranged during January 2018 and December 2017 respectively. The visits included four PD sessions that were held with 43 participants, including adult females and males, and teenage boys to provide a range of users input (teenage girls did not participate because they were having their pre-exam school break).

In Zaatari camp, the participants were chosen randomly from the resident workers and students of the camp's schools. Two sessions were held, involving 12 adult females and 12 teenage boys, respectively. The participants of each session were divided into three groups. The total of six designed mockups resulted from the sessions in Zaatari camp. However, one of the groups had to leave before completing their mockup. In the case of Azraq camp, the participants were chosen randomly from the records of an NGO. Two sessions were held. The first one involved 14 adult female participants who produced three mockups. A second session with a group of five adult male participants followed, and a plan of their preferred shelter was drawn. Three mockups and a 2D plan were the outputs of the experiment at Azraq camp.

Each session started with a five-minute introduction about the background of the researcher and the research, the purpose of the field visit, and an explanation of the experiment. The participants were informed that the discussions would be recorded and that photos of their work would be taken during the experiment. Each group was informed that the duration of the session was to be between 60 and 90 minutes, and they were made aware of their freedom to leave at any time and for any reason. Following the introduction, the participants were asked a few questions to initiate the discussions and to prepare them for the experiment. The questions were as follow:

— Before arriving to the camp, what did you expect your shelter to be like?
— What are the main challenges you face while living in the shelter?

The aim of asking these questions was to provoke the minds of the participants to answer the final question:

— What are the activities in which you engage in your daily life that require space inside your shelter?

The answers to the last question were written in lists for everyone to see during the experiment. Following the discussion, the participants were asked to split up into groups of four to

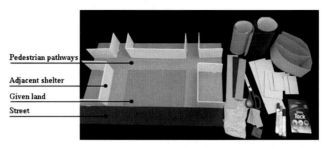

FIGURE 5.2   The toolkit of materials distributed to each group in the experiment. *From Alshawawreh L., Architecture of emergencies in the Middle East: Proposed shelter design criteria, (Unpublished Doctoral dissertation). 2019, Edinburgh Napier University, Edinburgh, United Kingdom, Retrieved from. http://researchrepository.napier.ac.uk/Output/2005914.*

six members, depending on the overall number of the participants in each session. The purpose was to design 3D mockups and/or plans for shelters that reflect the culture and functional preferences of the participants. Each group was given a set of prepared materials at the 1:25 scale, as listed below and shown in Fig. 5.2:

— A baseboard that included the layout of the given land, the layouts of the neighboring shelters, and the main street. The dimensions of the land plot were 7.5 m × 4.8 m (considering that each shelter would serve six people with a minimum area of 6 m^2 per person);
— Cardboards of different lengths and a "scaled dimension" height of 2.8 m. The cardboard lengths were marked every 1.2 m by a smooth slice, to help the participants in estimating the areas and the researcher in reading the mockups;
— Materials cut in the shape of doors (0.9 m × 1.8 m) and windows, in two sizes: small (0.5 m × 0.5 m) and big (1 m × 1 m);
— Other cardboard and EVA foam (soft polymer) pieces;
— A pack of putty removable adhesive, that is, blue tack;
— Scissors;
— Markers.

The participants were asked to consider the outlined plot on the board (Fig. 5.2) as a real plot of land that was given to them to build their own shelter. The provision of simple materials intended to ease and facilitate the interaction of the participants. The 3D mockups were then transformed into 2D plans by the researcher. However, some edits were made to the plans as the mockup dimensions were not accurate. Moreover, the doors were not drawn on the 2D plans as the type of doors was not discussed with the participants, but their location is indicated as openings.

The plans were then color-coded according to the level of privacy: blue for the public areas, red for the semiprivate areas, green for the private areas, and yellow for the facilities. The colors assisted in clarifying the use and circulation (movement between rooms). As the participants had no previous design experience, the resulting 3D mockups and plans were not ready designs. Instead, they were used to identify the residents' priorities, desired use of space, and functional needs. The resulting mockups were analyzed to extract the influencing factors in designing post-disaster shelters for the Syrian refugees. A summary of the sessions' details is illustrated in Table 5.1.

TABLE 5.1   Comparison between the sessions' details.

Camp		Azraq camp		Zaatari camp	
Sessions	**Session 1**	**Session 2**		**Session 1**	**Session 2**
Number of participants	14	5		12	12
Gender and age group of participants	Adult females	Adult males		Adult females	Teenage boys
Number of groups	3	3		3	1
Outputs	3 mockups	3 mockups: 2 finished and 1 unfinished mockup		3 mockups	2D plan

To sum up, the stages of the experiment were

— discussions with participants about their needs regarding their living spaces;
— the participants modeling their preferred shelters by using the distributed kits;
— transferring the models into 2D drawings by the researcher; and
— comparing the drawings and identifying patterns to take forward.

4. PD experiments at Zaatari and Azraq camps

Following the introduction, the PD experiments started with discussions in regard to the participants' experience in the camp, which included the functions they undertook in their shelters and their needs regarding space. Table 5.2 illustrates a comparison between the needs regarding space listed through the different sessions. Similar functions were repeatedly mentioned. However, some unique functions were also identified in certain sessions due to the difference in lived experiences among participants.

4.1 Shelter mockups at Zaatari camp

4.1.1 Zaatari camp: session 1

There were 12 female participants in this session who were divided during the experiment into three groups. During the discussions, the participants emphasized the importance of having an outdoor private area. The three designs included a courtyard, a family sitting room, two bedrooms, a kitchen, a toilet, and a shower. Fig. 5.3 shows the three mockups, and Fig. 5.4 shows the 2D plans that represent the mockups.

The first and second groups specified a reception area for the guests, while the third group did not include a reception area but instead considered the courtyard as a reception area. The third group provided an unlabeled room that is assumed to be a toilet for guests, as it could be accessed through the courtyard. The first group had the entrance from the street side and directly into the sitting room, which distributes to most other functions including the court-yard as shown in Fig. 5.4A. This circulation gives guests more accessibility to the family

TABLE 5.2 Comparison between the participants' proposed function needs during the session.

Suggested function	Zaatari camp		Azraq camp	
	Session 1: Adult females	Session 2: Teenage boys	Session 1: Adult females	Session 2: Adult males
Private courtyard	✔	✔	✔	✔
Private garden	✔		✔	✔
Three bedrooms	✔	✔	✔	✔
Kids room (playing and studying)		✔	✔	✔
Reception	✔	✔	✔	✔
Family sitting area	✔	✔	✔	✔
Storage area			✔	✔
Kitchen	✔	✔	✔	✔
Separate toilet and shower	✔	✔	✔	✔
Private water tank			✔	✔
Continuous floor			✔	✔
Outdoor fence	✔			

sitting room and less accessibility to the courtyard. Having a bedroom access through the kitchen is not an ideal situation, but it was concluded that it was a solution to a space problem faced by group 1 while developing their mockup. The second and third groups had their entrances from the side of the shelters and directly into the courtyards. The second group had two secondary entrances as shown in Fig. 5.4B, one for family members that leads into a family sitting room conducting to all other functions, and a second for visitors in the reception room. The third group made an entrance from the courtyard into the family sitting room (Fig. 5.4C). The circulation in the second and third groups' mockups starts from a public area (courtyard) into a semiprivate area (family sitting room) and then into the private areas (bedrooms).

Although all the participants preferred to have a toilet that is separated from the shower, this was not possible due to space limitations. The entrance to the shower in all three designs was through the toilet but subdivided by walls. The participants positioned the windows toward the private courtyards whenever possible, while other windows were inserted in most rooms and positioned as high as possible to limit the visibility from outside. In addition, the participants mentioned the importance of fencing the shelters to enhance privacy. Group 2, in Fig. 5.4B, and group 3, in Fig. 5.4C, introduced the use of inner windows in their designs to enhance the air movement, reproducing a feature that was implemented in their previous homes in Syria.

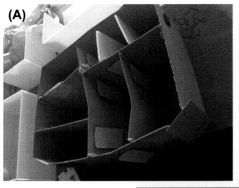
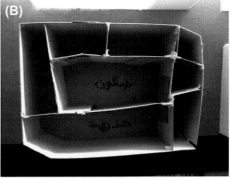

FIGURE 5.3 The 3D mockups designed by Zaatari-camp residents during the first session with females: (A) proposed design by group 1, (B) proposed design by group 2, and (C) proposed design by group 3. *From Alshawawreh L., Architecture of emergencies in the Middle East: Proposed shelter design criteria, (Unpublished Doctoral dissertation). 2019, Edinburgh Napier University, Edinburgh, United Kingdom, Retrieved from. http://researchrepository. napier.ac.uk/Output/2005914.*

4.1.2 Zaatari camp: session 2

The second session involved 12 teenage boys, with ages ranging between 14 and 16 years old. The boys were divided into three groups of four participants each. The resulting mockups are shown in Fig. 5.5. The second group could not finish their mockup as they had school commitments. However, the mockups that were made by the first and third groups were transformed into 2D plans as shown in Fig. 5.6.

Both designs had their entrances directly from the street into the interior of the shelter. The teenage boys did not allocate courtyards in their design, which may reflect the fact that males do not spend as much time in the courtyards as females do. The two designs had a reception room for guests, three bedrooms, a kitchen, a toilet, and a shower. As shown in Fig. 5.6A, the first group had a family sitting room as the main distributor to the other functions and included a study room. The reception entrance in the design of the third group was inserted close to the main door as shown in Fig. 5.6B, to limit the movement of the guests throughout the shelter. The third group also preferred to dispense the courtyard with a third bedroom. The designs of the first and third groups separated the toilet from the shower. Similar to the female groups in Zaatari camp, the teenage boys positioned the windows on a high level. In terms of circulation, the design of the first group was not led by privacy concerns, unlike

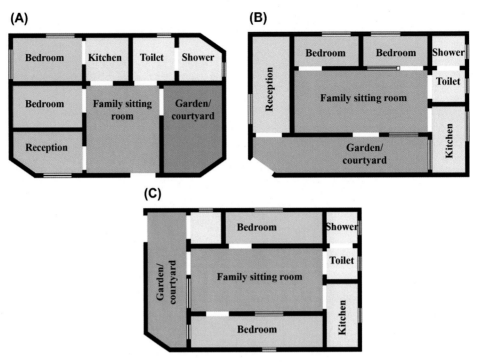

FIGURE 5.4 The 2D plans drawn from the 3D mockups produced through the first session in Zaatari camp: (A) the shelter plan by group 1, (B) the shelter plan by group 2, and (C) the shelter plan by group 3. Key: blue: public areas; red: semiprivate areas; green: private areas; yellow: facilities. *From Alshawawreh L., Architecture of emergencies in the Middle East: Proposed shelter design criteria, (Unpublished Doctoral dissertation). 2019, Edinburgh Napier University, Edinburgh, United Kingdom, Retrieved from. http://researchrepository.napier.ac.uk/Output/2005914.*

the third group who had a clear hierarchy of privacy from public areas to semiprivate and finally private areas.

4.2 Shelter mockups at Azraq camp

4.2.1 Azraq camp: session 1

There were 14 female participants in this session who were divided into three groups. The final mockups are shown in Fig. 5.7. All three groups have located their entrances off the main street directly into a private courtyard. The first group, as shown in Fig. 5.7A, made the access point to the interior through the reception room that leads into the family sitting room and then to the other functions. However, the second and the third groups, as illustrated in Fig. 5.7B—C, made separated entrances to most of the functions directly from the courtyard. The three groups included outdoor elements to identify the entrances. Moreover, the three designs had a courtyard, two bedrooms, a family sitting room, a toilet, and a kitchen, as shown in Fig. 5.8.

A reception room for guests was only proposed in the designs of the first and third groups. In terms of the sanitary facilities, the second group combined the shower and toilet in the same space (Fig. 5.8B), while the third group positioned the toilet next to the reception to limit

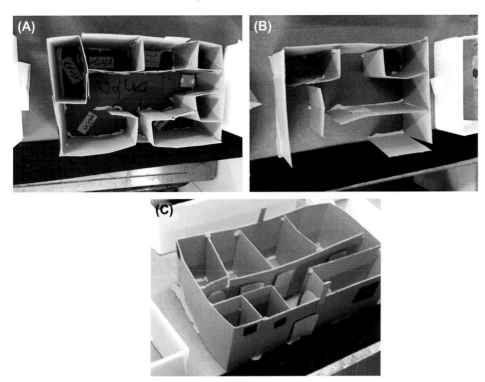

FIGURE 5.5 The 3D mockups designed by Zaatari-camp residents during the second session with teenage boys: (A) proposed design by group 1, (B) proposed design by group 2, and (C) proposed design by group 3. *From Alshawawreh L., Architecture of emergencies in the Middle East: Proposed shelter design criteria, (Unpublished Doctoral dissertation). 2019, Edinburgh Napier University, Edinburgh, United Kingdom, Retrieved from. http://researchrepository. napier.ac.uk/Output/2005914.*

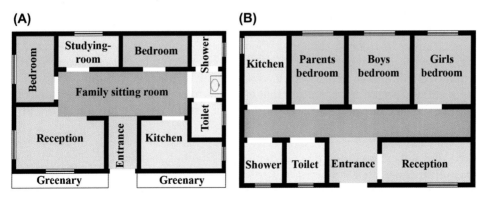

FIGURE 5.6 The 2D plans drawn from the 3D mockups produced through the second session in Zaatari camp: (A) the shelter plan by group 1 and (B) the shelter plan by group 3. Key: blue: public areas; red: semiprivate areas; green: private areas; yellow: facilities. *From Alshawawreh L., Architecture of emergencies in the Middle East: Proposed shelter design criteria, (Unpublished Doctoral dissertation). 2019, Edinburgh Napier University, Edinburgh, United Kingdom, Retrieved from. http://researchrepository.napier.ac.uk/Output/2005914.*

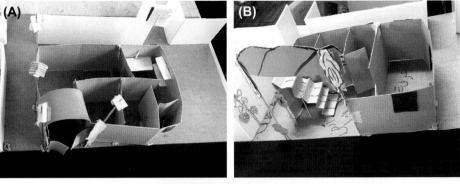

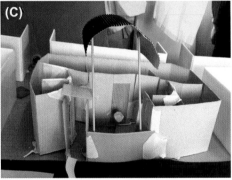

FIGURE 5.7 The 3D mockups designed by Azraq-camp residents during the first session with females: (A) proposed design by group 1, (B) proposed design by group 2, and (C) proposed design by group 3. *From Alshawawreh L., Architecture of emergencies in the Middle East: Proposed shelter design criteria, (Unpublished Doctoral dissertation). 2019, Edinburgh Napier University, Edinburgh, United Kingdom, Retrieved from. http://researchrepository. napier.ac.uk/Output/2005914.*

the movement of the guests, as shown in Fig. 5.8C. It was identified that the circulation between the rooms was based on privacy needs, that is, from public areas into semiprivate and then into private areas. The participants clarified that they prefer large family sitting rooms over larger bedrooms, similar to their original homes. In the three designs, most of the rooms had allocated windows, but some rooms were left windowless. Moreover, the windows that overlook public areas were located as high as possible to limit the visibility from outside, a cultural architectural feature that is also recalled from the refugees' home of origin. The participants also mentioned the importance of fencing the shelter to increase privacy.

4.2.2 Azraq camp: session 2

The second session included five resident men. Due to a logistical problem, the male participants represented their shelter design preferences through a 2D drawing instead of the usual 3D mockup. The participants mentioned issues in relation to the current distribution system of the shelters in the camp. They believed that the criteria should depend on the age and gender of the family members and not only on their number. Fig. 5.9 shows the participants' proposed plan next to a color-coded edited version. As the template design kit was not used, the resulting plan had unrealistic building dimensions. Therefore, the plan was rationalized to fit in the same outline that was given to the other groups.

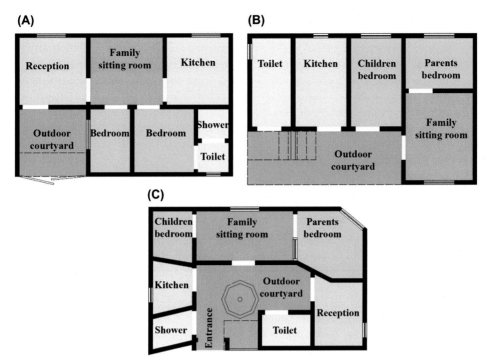

FIGURE 5.8 The 2D plans drawn from the 3D mockups produced through the first session in Azraq camp: (A) the shelter plan by group 1, (B) the shelter plan by group 2, and (C) the shelter plan by group 3. Key: blue: public areas; red: semiprivate areas; green: private areas; yellow: facilities. *From Alshawawreh L., Architecture of emergencies in the Middle East: Proposed shelter design criteria, (Unpublished Doctoral dissertation). 2019, Edinburgh Napier University, Edinburgh, United Kingdom, Retrieved from. http://researchrepository.napier.ac.uk/Output/2005914.*

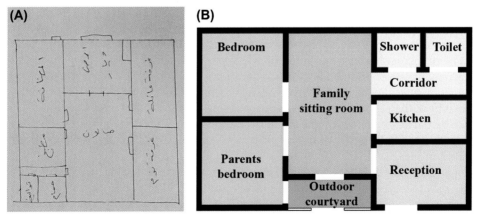

FIGURE 5.9 The 2D plan produced through the second session in Azraq camp with males: (A) the hand-drawn plan and (B) the edited shelter plan fitted into the standard area. Key: blue: public areas; red: semiprivate areas; green: private areas; yellow: facilities. *From Alshawawreh L., Architecture of emergencies in the Middle East: Proposed shelter design criteria, (Unpublished Doctoral dissertation). 2019, Edinburgh Napier University, Edinburgh, United Kingdom, Retrieved from. http://researchrepository.napier.ac.uk/Output/2005914.*

The male group did not distribute any windows in their design, but they verbally expressed their concerns regarding the windows that overlook the neighboring shelters or the public areas. However, overlooking the public is unavoidable in their plan, if any window is to be inserted. The privacy concerns were the main leading element in the space planning of this design. The public courtyard can be accessed from the street, and the semiprivate area (family sitting room) provided the main distribution to all other areas. This group separated the shower from the toilet and made them open into a small corridor. The uniqueness of this design lies in having a reception room that has two entrances: one from the street side and the other from the family sitting room. The two entrances aimed at limiting the accessibility of the guests to the interior of the shelter.

5. Discussion

The functions inside the shelters were common to most designs as they included an outdoor courtyard, a reception, a family sitting room, two bedrooms, a kitchen, a toilet, and a shower. Some designs had extra functions such as an additional room, while others had less functions and lacked the courtyard or the reception.

The aforementioned fact about the origin of the Syrians in Jordan being mainly from rural areas justifies some of the proposed designs. For instance, the importance given to the courtyard comes mainly from the typical rural houses. According to CORPUS Levant (2004), who analyzes the traditional architecture in Syria, the lifestyle of the people who live in rural areas and depend on agriculture and livestock farming requires the house to have an attached courtyard. Another example is the separation of the toilet from the shower and the transfer of the toilet outside the shelter, which also originates from the typical design of rural housing. The limited number of openings in the shelter that aimed at increasing the privacy is also an approach commonly endorsed in Syrian rural architecture (CORPUS Levant, 2004). Thus, we can conclude that socioeconomic and cultural factors strongly influenced the residents' choices in their design of their ideal shelter.

As for the difference in design approaches between the two studied camps, the only noticeable contrast was the relation between private courtyards and public surroundings. In Azraq camp, the participants chose light materials to separate the courtyards, while in Zaatari camp, the participants surrounded the courtyards with solid walls. This preference may refer to their current safety experience. However, it was noted that living in different shelter types did not affect their perspective and approach regarding the design of their ideal shelter. This can be partly explained by the similarity of the one-room shelters from a spatial perspective in both camps. In terms of the participants' gender, some clear differences were found, illustrated in Fig. 5.10.

The gender-related design differences consisted in the courtyard size, the number and position of entrances, the size and position of windows, and the number of rooms. In addition, the connections between the functions differed, notably between the toilet and the shower, the toilet and the reception, and the courtyard and the reception. The size of the courtyard was a major difference, as females designed larger courtyards, an evidence of the importance of this space for them, given the significant amount of time they spend inside shelters. Males

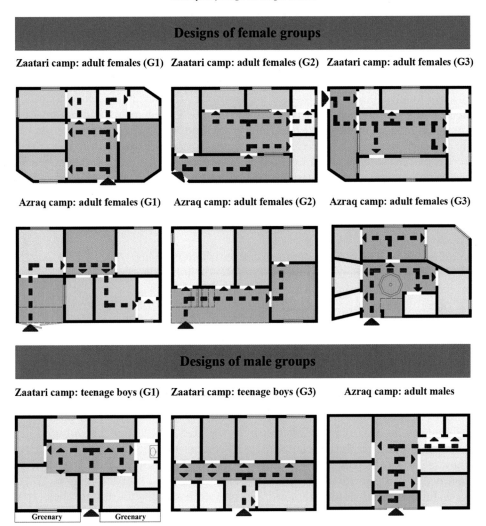

FIGURE 5.10 Comparing the nine shelter designs. Key: blue: public areas; red: semiprivate areas; green: private areas; yellow: facilities. *From Alshawawreh L., Architecture of emergencies in the Middle East: Proposed shelter design criteria, (Unpublished Doctoral dissertation). 2019, Edinburgh Napier University, Edinburgh, United Kingdom, Retrieved from. http://researchrepository.napier.ac.uk/Output/2005914.*

made smaller or no courtyards and preferred to have larger interior spaces. The freedom of movement for males inside the camps may be the incentive for their choice.

Another difference is the number and position of entrances. The male group preferred two shelter entrances: one for family members and a separate entrance for the guests, increasing the shelter's privacy based on cultural motives. Conversely, all of the female groups designed one entrance, with five out of the six groups adding secondary entrances throughout the courtyards. The three male groups designed entrances from the street side, leading directly into the middle of the shelter, but the female groups achieved the needed privacy through making the main entrance open into the courtyard, and therefore away from the private areas, as clarified in Fig. 5.10.

Moreover, the female groups inserted windows that overlook the private courtyard, but also inserted high-level windows in other external walls, and recommended to have outdoor solid fences for more privacy. The two groups of teenage boys did not include windows in all rooms, while the adult males (who did not position the windows in their plan) stated verbally their refusal to have windows overlooking streets or neighbors.

Regarding toilets, although all the groups asserted in the discussion phase that they prefer the toilet and the shower to be in separate rooms, some female groups did not prioritize that, contrary to the males. On the other hand, the male groups appeared not to consider the relationship between the toilet and reception, contrary to some female groups who positioned these two functions side by side to preserve the privacy for the rest of the shelter.

The relation between the courtyard and the reception was also approached differently. The female groups located the main entrance through the courtyard, which is accessible to guests, while the group of adult males created separate outdoor entrances to the reception. Other gender-related design differences include having larger family sitting rooms according to the designs of the males.

The aforementioned existing literature on the shelters in the Zaatari and Azraq Syrian camps have covered the thermal inadequacy of the shelters (Albadra et al.,. 2018), highlighted the drawbacks caused by the choice of the existing shelters, such as the lack of proper floor and the rusted sandwich panels (ACTED, 2017; REACH, 2014, 2015), and emphasized the lack of cultural consideration and shortage of extension possibilities (Alshawawreh et al., 2017). Based on these findings, this chapter demonstrates the importance of engaging the users in the design process.

The previously mentioned cases of using PD experiments to design shelters in Ecuador and Kenya provided kits of ready-modeled rooms to the participants to choose from, which were stated to be results of previous drawing exercises (Frediani et al., 2013). However, the ready-modeled rooms used in these experiments limited the choices of participants and hence the fulfillment of their creativity. This research provided an additional stratum in PD experiments and in studies on Syrian camps. The researcher engaged the residents in designing what they see as ideal shelters by proposing a toolkit that allowed them to express undeniably their needs and priorities. The challenge that was argued by Sanders et al. (2010) about finding techniques that could be used by nonprofessionals to design their space was definitely met in this research.

6. Conclusion

The aim of this participatory design experiment was not to have ready designs to take forward, but to understand how the residents would approach the shelter design, to know their priorities, and based on this information to be able to provide assistance for future shelter designs. Given the increased numbers of protracted refugees in the world (UNHCR, 2019c), this study shows the importance of engaging the users in the shelter designs to duly meet their needs and cultural aspirations, while also empowering them by valuing their voice.

Each design had its own identity, and the participants' diversity enriched the outcomes of the study. But at the same time, there were many commonalities that came from the cultural identity and beliefs of the community members. The participants were inspired by the building typology of the Syrian traditional architecture. The two main factors that affected the participants' decisions were culture and privacy. For most of them, considering age and gender factors when distributing the shelters was an essential request. Additionally, there were clear differences between male and female designs in terms of layout, space requirements, and functional needs. For instance, the outdoor private courtyard/garden was highlighted as one of the most important spaces, especially for females due to the significant amount of time they spend inside their shelters.

The circulation within the shelter is an example of the cultural influence, as it was of significant concern to design it in a way that would protect the inhabitants' privacy. In fact, the movement between "public," "semiprivate," and "private" areas within the shelter was an important concern to most participants. Inside the plot plan, the family sitting area turned out to be the most important space that acts as a distributor to other rooms. Participants favored windows that overlook private areas, but whenever this was not possible, they positioned them in their designs at high levels, and projected the construction of a solid fence surrounding the shelter. The participants also preferred to have inner windows connecting different rooms to the sitting area to intensify the air movement. Flexibility in the design is a significant feature that provides the users with the opportunity to create more individuality, generating a sense of belonging and enhancing their wellbeing. Such processes obviously require increased levels of planning and coordination.

Failure to address the refugees' needs in the initial shelter designs encourages them to make unplanned changes. This could lower the quality of the shelters and at the same time impact the urban landscape. Unplanned changes cause the evolvement of the settlement or the camp in a series of irregular constructions with sporadic additions from a design perspective, and these would end up turning into slums..

The research had some limitations, such as the geographic distance, the limited access to the camps, and the sensitivity of the situation inside the camps. These factors restricted the sample numbers and their demographic structure. However, the difference in age and gender of the participants contributed to having more balanced findings. Future research could include an organized series of PD sessions involving a larger number of participants with various demographic characteristics. They would take place in different contexts, hopefully leading to more inclusive future shelter designs.

Acknowledgments

The author would like to thank Plan International and Save the Children in Jordan for organizing the visits to Azraq camp and Zaatari camp. The staff in both organizations was extremely supportive and generously offered their help and assistance. Additionally, grateful thanks go to Prof Sean Smith, who enriched the analysis stage of this research by his insights and guidance.

References

ACTED. (2017). Shelter intervention snapshot. *Jordan: Za'atari refugee camp.* Retrieved from https://data2.unhcr.org/en/documents/download/61479.

Albadra, D., Coley, D., & Hart, J. (2018). Toward healthy housing for the displaced. *The Journal of Architecture, 23*(1), 115–136. https://doi.org/10.1080/13602365.2018.1424227.

Alshawawreh, L. (2019). *Architecture of emergencies in the Middle East: Proposed shelter design criteria* (Unpublished Doctoral dissertation). Edinburgh, United Kingdom: Edinburgh Napier University. Retrieved from http://researchrepository.napier.ac.uk/Output/2005914.

Alshawawreh, L., Smith, R. S., & Wood, J. B. (2017). Assessing the sheltering response in the Middle East: Studying Syrian camps in Jordan. *International Journal of Humanities and Social Sciences, 11*(8), 2034–2040. https://doi.org/10.5281/zenodo.1131551.

CARE International. (2015). *Azraq refugee camp, Jordan.* Factsheet April 2015. Retrieved from https://www.care.org/sites/default/files/documents/CARE_Factsheet_Syria_azraq_camp_April_2015.pdf.

Carroll, J. M. (2006). Dimensions of participation in Simon's design. *Design Issues, 22*(2), 3–18. https://doi.org/10.1162/desi.2006.22.2.3.

CORPUS Levant. (2004). *Traditional Syrian architecture.* Avignon: École d'Avignon.

Devictor, X., & Do, Q.-T. (2016). *How many years have refugees been in exile?.* World Bank Policy Research Working Paper 7810. Retrieved from http://documents.worldbank.org/curated/en/549261472764700982/pdf/WPS7810.pdf.

Frediani, A. A., De Carli, B., Ferrera, I. N., & Shinkins, N. (2013). *Change by design: New spatial imaginations for Los Pinos.* Architecture Sans Frontières UK Workshop report. Retrieved from http://www.asf-uk.org/wp-content/uploads/2016/05/CbD2013_LP-report-EN-Download.pdf.

French, M. A. (Ed.). (2011). *Change by design: Building communities through participatory design.* Napier, New Zealand: Urban Culture Press.

Gatter, M. N. (2018). Rethinking the lessons from Za'atari refugee camp. *Forced Migration Review, 57,* 22–24. https://doi.org/10.17863/CAM.22910.

Gavlak, D. (January 8, 2013). *Winter storm brings more misery to Syrian refugees.* Retrieved from https://news.yahoo.com/winter-storm-brings-more-misery-syrian-refugees-193635388.html.

Global Shelter Cluster. (2017). *Shelter projects 2015–2016.* Retrieved from http://shelterprojects.org/shelterprojects2015-2016/ShelterProjects_2015-2016_lowres_web.pdf.

IFRC. (2013). *Post-disaster shelter: Ten designs.* Retrieved from https://www.sheltercluster.org/resources/documents/post-disaster-shelter-ten-designs.

IFRC, UN-Habitat, & UNHCR. (2014). *Shelter projects 2013–2014.* Retrieved from http://shelterprojects.org/shelterprojects2013-2014.html.

International Labour Organization. (October 4, 2017). *EU-funded Zaatari camp job centre hosts unprecedented job fair.* Retrieved from https://www.ilo.org/beirut/media-centre/news/WCMS_579487/lang–en/index.htm.

Kennedy, J., & Parrack, C. (2013). The history of three point five square metres. In IFRC, UN-Habitat, & UNHCR (Eds.), *Shelter projects 2011–2012* (pp. 109–111). Retrieved from http://shelterprojects.org/shelterprojects2011-2012/B01-3point5.pdf.

Ledwith, A. (2014). *Zaatari: The instant city.* Boston, MA: Affordable Housing Institute.

Maayeh, S. (January 7, 2015). Syrian refugees die in worst Middle East storm in decades. *The National.* Retrieved from https://www.thenational.ae.

Ministry of Planning and International Cooperation. (2017). *Jordan Response Plan for the Syria Crisis 2017–2019. 2017 annual report.* Retrieved from https://reliefweb.int/sites/reliefweb.int/files/resources/JRP%2B2017-2019%2B-%2BFull%2B-%2B%28June%2B30%29.pdf.

REACH. (2014). *Al Za'atari refugee camp shelter assessment.* Retrieved from https://data2.unhcr.org/en/documents/download/70965.

REACH. (2015). *Azraq camp shelter assessment.* Retrieved from https://www.impact-repository.org/document/reach/6324449e/reach_jor_report_azraqcampshelterassessment_jan2015.pdf.

Roth, S. (1999). The state of design research. *Design Issues, 15*(2), 18–26. https://doi.org/10.2307/1511839.

Sanders, E. B.-N., Brandt, E., & Binder, T. (2010). A framework for organizing the tools and techniques of participatory design. In T. Robertson (Ed.), *Proceedings of the 11th Biennial Participatory Design Conference* (pp. 195–198). New York, NY: ACM. https://doi.org/10.1145/1900441.1900476.

Siegel, R. (March 17, 2017). *UN High Commissioner for Human Rights calls Syria a 'torture chamber'.* Retrieved from https://www.npr.org/2017/03/17/520576917/u-n-high-commissioner-for-human-rights-calls-syria-a-torture-chamber.

Sphere Association. (2018). *The Sphere Handbook: Humanitarian Charter and minimum standards in humanitarian response* (2018 ed.). Retrieved from http://www.spherestandards.org/handbook/editions.

Sphere Project. (2011). *Humanitarian Charter and minimum standards in humanitarian response.* Retrieved from http://www.spherestandards.org/handbook/editions.

Tiltnes, Å. A., Zhang, H., & Pedersen, J. (2019). *The living conditions of Syrian refugees in Jordan.* Fafo report 2019:04. Oslo: Fafo. Retrieved from http://dosweb.dos.gov.jo/DataBank/Analytical_Reports/English/Syrian_refugees.

UNHCR. (2015). *Transitional shelter (T-shelter) design for Azraq camp, Jordan.* Retrieved from https://data2.unhcr.org/en/documents/download/46026.

UNHCR. (2016). *Shelter design catalogue.* Retrieved from https://cms.emergency.unhcr.org/documents/11982/57181/Shelter+Design+Catalogue+January+2016/a891fdb2-4ef9-42d9-bf0f-c12002b3652e.

UNHCR. (2018a). *Global trends. Forced displacement in 2017.* Retrieved from http://www.unhcr.org/5b27be547.pdf.

UNHCR. (2018b). *Jordan: Zaatari camp, Fact sheet February 2018.* Retrieved from https://data2.unhcr.org/en/documents/download/62238.

UNHCR. (2019a). *Emergency handbook: Site planning for camps.* Retrieved from https://emergency.unhcr.org/entry/254415/site-planning-for-camps.

UNHCR. (2019b). *Figures at a glance.* Retrieved from https://www.unhcr.org/figures-at-a-glance.html.

UNHCR. (2019c). *Global trends: Forced displacement in 2018.* Retrieved from https://www.unhcr.org/5d08d7ee7.pdf.

UNHCR. (2019d). *Jordan: Azraq refugee camp, Fact sheet August 2019.* Retrieved from https://data2.unhcr.org/ar/documents/download/71530.

UNHCR. (2019e). *Jordan: Azraq refugee camp, Fact sheet January 2019.* Retrieved from http://reporting.unhcr.org/sites/default/files/UNHCR%20Jordan%20Azraq%20Refugee%20Camp%20Fact%20Sheet%20-%20January%202019.pdf.

UNHCR. (2019f). *Syria regional refugee response.* Retrieved from https://data2.unhcr.org/en/situations/syria.

USA for UNHCR. (July 17, 2017). *Making the Za'atari refugee camp a community.* Retrieved from https://www.unrefugees.org/news/making-the-za-atari-refugee-camp-a-community.

6

Lessons for humanitarian architecture from design contests: The case of the Building 4Humanity Design Competition

A. Nuno Martins[1], Liliane Hobeica[2], Adib Hobeica[3], Raquel Colacios[4]

[1]CIAUD, Research Centre for Architecture, Urbanism and Design, Faculty of Architecture, University of Lisbon, Lisbon, Portugal; [2]RISKam (Research group on Environmental Hazard and Risk Assessment and Management), Centre for Geographical Studies, University of Lisbon, Lisbon, Portugal; [3]Independent consultant, Coimbra, Portugal; [4]School of Architecture, Universitat Internacional de Catalunya, Barcelona, Spain

Enhancing Disaster Preparedness
https://doi.org/10.1016/B978-0-12-819078-4.00006-X

1. Introduction

Over the past decade, humanitarian architecture (HA) has evolved from a socially engaged design practice to an emerging body of knowledge (Aquilino, 2011; Charlesworth, 2014; Wagemann & Ramage, 2013). In a world with increasing frequency and impacts of disasters and conflicts, and with prevailing inequalities, architects play an important role in recovery and development processes. This is for instance exemplified by the work of NGOs and individual professionals who have been contributing to post-disaster interventions and development-related architectural practices, that is, building and design in development scenarios, such as resettlement and slum-upgrading projects (Wagemann & Ramage, 2013). However, the contexts in which HA is called to intervene always pose challenges that are usually not covered in traditional architectural training (Acar & Yalçınkaya, 2016). Indeed, as "architects are taught to focus on the product (a building), whereas humanitarian practitioners major on the process (involving people)" (Sanderson, 2010), the architects who choose to work in the development or the humanitarian-aid spheres often have to learn by doing, while reframing some professional standards.

Given the prominence of poverty and unplanned urban expansions in the Global South, architectural and urban-planning education has gradually been mainstreaming development-related issues in undergraduate programs in some of these countries. This is illustrated by academic experiences in Brazil (Benetti & Carvalho, 2017), India (Narayanan, 2013), Thailand (Tovivich, 2010), among others. Conversely, disaster risk reduction (DRR) is most often absent from undergraduate programs (Acar & Yalçınkaya, 2016; Brogden, 2019; Wagemann & Ramage, 2013). Post-disaster architectural topics are presently only covered in a few graduate courses created in the last three decades (Aquilino, 2011; Charlesworth, 2015; Ehrmann, 2018). For instance, the Centre for Development and Emergency Practice (CENDEP, Oxford Brookes University, Oxford, UK), the Royal Melbourne Institute of Technology (RMIT, Melbourne, Australia), the Bartlett Development Planning Unit (University College London, London, UK), and Universitat Internacional de Catalunya (UIC, Barcelona, Spain) are among the institutions that presently offer specialized training related to disaster reconstruction, risk prevention, resilience, development, and design. These programs seek to train architects and other built-environment professionals with new and interdisciplinary skills needed to work in post-disaster and vulnerable contexts.

Considering that the humanitarian-architecture practice has been growing in relevance while being revised and reoriented, there remains an open question: how is architectural education dealing with such a domain? It thus seems necessary to include humanitarian architecture's concepts, theories, and experiences in the education sphere. This move would discourage traditional architectural practices from negatively interfering with the transformation processes of communities at risk or affected by disasters. At the same time, it would allow architects to be trained to deal with the complexity of development-related and post-disaster reconstruction processes, expanding their skills to effectively play their expected roles (see Chapter 2, by Andriessen, Paidakaki, Susilo, and Van den Broeck, in this volume).

This chapter examines international design competitions that contribute to bridge the gap between architectural training and the required knowledge for post-disaster reconstruction and recovery. Our study intended to understand how design competitions can support humanitarian-architecture training and contribute to its consolidation. After reviewing key literature in Section 2, we explore in Section 3 the trajectory of two design competitions focusing on risk and resilience: i-Rec (Information and Research for Reconstruction) and DRIA (Designing Resilience in Asia). We then analyze in detail the outcomes of the Building 4Humanity Design Competition (B4H-DC), organized by the Portuguese NGO Building 4Humanity Designing and Reconstructing Communities Association (B4H). In the discussion and conclusion sections, we highlight some lessons learned from the B4H-DC and their major implications for architectural education and practice.

2. Mainstays of humanitarian architecture and the role played by architectural education

Humanitarian architecture has usually been understood as the work of architects in emergency contexts framed by conflicts and disasters (Aquilino, 2011; Charlesworth, 2014). Other authors, such as Sinclair (2012) and Martins and Rocha (2019), took a more wide-ranging approach and included the work of architects with deprived communities living in slums or impoverished areas, for instance, as part of HA. Yet, the demands of post-disaster and conflict settings and the ones of poverty- and exclusion-stricken contexts are very different in several aspects, such as timeframe, resources involved, level of engagement and participation, partnerships sought, and funding. Moreover, while in the former architectural practices must quickly respond to short-term and emergency needs, in the latter the involved issues relate to longer-term development endeavors. Nonetheless, in both cases architects are required to act differently from the way the profession is commonly taught in architecture schools worldwide (Sanderson, 2010).

This point was emphasized by Brogden (2019, p. 2), for whom "humanitarian and development projects are increasingly calling for more participatory, community-led approaches in which the role of the architect is necessarily that of a 'humble facilitator'." When indicating the various fields in which the HA approach is currently required, Charlesworth (2014) also included a broader array of projects addressing vulnerable communities, such as slum upgrading or housing for internally displaced people. Indeed, for such groups, the occurrence of disasters may be an imminent threat.

In her attempt to define HA, Brogden (2019, p. 3) proposed the descriptor "design for crisis situations," though underlining the difficulty in reaching a universal agreement on what this category is about. On the other hand, Martins and Guedes (2015) explored the differences between conventional and humanitarian approaches to architecture and outlined a set of principles for a HA practice. Further pursuing a clarification of the HA concept, Martins and Rocha (2019) analyzed the tools used in the projects conducted by NGOs in Brazil and Africa. Given the positive results of the collaboration between different specialists, they argued for a transdisciplinary approach that expands the full spectrum of architecture while recognizing the importance of fostering translational tools to improve the communication among professionals, and between them and other stakeholders. Eventually, these authors argued that DRR, vulnerable minorities, gender issues, and social-innovation tools—namely the use of collective mapping and collaborative design with local communities—should be taken as key mainstays of HA practice (Martins & Guedes, 2015; Martins & Rocha, 2019).

HA requires the abovementioned design pillars to face the complex challenges that its practice implies. In that direction, Boano and Hunter (2012) recognized that DRR-related architectural outputs are concomitantly compromised in the spatial, social, and political dimensions. These authors argued that hasty post-disaster rebuilding actions might impend the architectural quality and the appropriation of future preventive measures. They also claimed that post-disaster reconstruction often erodes communities, livelihoods, and security, nurturing social fragmentation. Indeed, reconstruction is a process that expands to political and institutional agendas in which time, money, and infrastructures create a tense environment marked by a chase to deliver as much and as fast as possible. Such a scenario usually clashes with social demands and aspirations, and breaks everyday practices and place attachment, especially when resettlements and nonculturally appropriate physical solutions are involved (Boano & Hunter, 2012).

Yet, some authors have debated the positive roles that architects play in both disaster preparedness and post-disaster reconstruction, such as "translators," "mediators," "cultivators," or "facilitators" (Boano & García, 2011; Brogden, 2019; Martins & Rocha, 2019; Tauber, 2014). According to Lizarralde, Johnson, and Davidson (2010), the remit of architects proceeds from the interpretation of the different capacities of stakeholders rather than from their traditional design and construction tasks. These authors argued for the reshaping of the role played by architects in emergency response and humanitarian-crisis situations in general, toward an enabling intermediary role. In contrast, we take the perspective of architects' role as both practical and artistic, and each piece of good architectural work as a unique creation, "a mediation between our world and our minds," in the words of Pallasmaa (2018).

Yet, these different views do not necessarily have to be seen as antagonistic. Prominent theoreticians such as Bachelard (1957/1994) and Norberg-Schulz (1980), or practitioners such as Zumthor (2006) or Pallasmaa (2014), converge to distinguish in architecture the very foundations of the human soul and a given culture. The ability of architectural practices to meet people's aspirations through imagination (Hamdi, 2004), providing for more than the immediate emergency or primary needs (Davis & Alexander, 2016), can be shared by both educators and professionals.

Indeed, a strong credit for a humanitarian approach to architecture can be found in those who acclaim an increased social approach to architecture in crisis scenarios, as illustrated by Lepik's 2010 MoMA exhibition *Small scale, big change* (Lepik, 2010). This is also recognizable in the claims for architects' expanded role in servicing and representing the addressed communities. For instance, Aquilino (2011) stood for artistic creativity in architectural recovery projects. This creativity is undoubtedly one of the main features acknowledged in the works of Shigeru Ban and Alejandro Aravena, two Pritzker Architecture Prize awardees in the past six years for their outstanding humanitarian-architecture contributions. Disaster-related interventions and the action-research arena provide other expressive examples of confluence between practitioners and theoreticians. The architecture of risk, referred to by the architect and NGO worker Victoria Harris (2011), or the architecture of development, as defined by the champion of urban participatory design, the architect and educator Nabeel Hamdi (2004), are strongly convergent. They coincide in stressing vulnerabilities and the need for doing more (for people's safety and wellbeing) with less (resources) while investing in small urban changes (at the microscale) that may have future multiplier effects (onto the entire urban setting).

In their introduction of *Rebuilding after disasters*, Lizarralde et al. (2010, p. 2) distinguished the "common misconceptions or myths from factual realities of reconstruction." One of these myths is "that community participation holds the key to successful reconstruction" (Lizarralde et al., 2010, p. 2). They alerted to side effects and negative long-term results of short-term community decisions and the need for ensuring, during the participation process, "real decision-making power over design, planning, financing and management of the projects for individual users" (Lizarralde et al., 2010, p. 13). These scholars further suggested that it is the responsibility of building professionals to "interpret the ways of living of affected residents and the housing typologies of disaster-affected areas" and "design solutions capable of promoting long-term development" (Lizarralde et al., 2010, p. 24). They concluded by pointing out "the absence of competent decision-makers and designers" as one of the causes of the "development of inappropriate solutions that are—often—as dangerous and problematic as the disaster itself" (Lizarralde et al., 2010, p. 24). In tune with this view, in *Beyond shelter*, although from a perspective more related to architects' role, Aquilino (2011) pointed to the significant limitations of communities regarding capacity, representation, and vision.

To fill the gap between design and participation, or between architecture and people's short- and long-term needs, architects have always endeavored to know, listen to, understand, and reverberate the communities and places they address in every project. Moreover, architecture has consistently reinvented and consolidated itself through sounding concepts that are rooted in its theory and practice. We recall here some of these. First, building typology, as revisited in the 1950s by Muratori (1960) to study the history of old-town areas of Italian medieval cities and in the 1980s by Moneo (Durand, 1981) in his prologue to the Spanish edition of Durand's *Précis des leçons d'architecture données à l'École Royale Polytechnique*. Second, place-making and the spirit of the place (or *genius loci*), stressed by Norberg-Schulz (1980) in his attempt of a phenomenological interpretation of architecture, speak to the concern of matching communities' physical and psychological priorities while enhancing their connections with their environments. Third, atmospheres, as described by Pallasmaa (2014) and Zumthor (2006), who understand this notion as the architectural embodiment of the sense of the place, in Norberg-Schulz's terms, highlight the power of architecture to convoke all our sensorial perceptions. Thus, atmospheres aim at making real the vision of a better future, which is actually a critical function of architecture in postcrisis situations, as remarked by Aquilino (2011).

Authors such as Aquilino (2011), Charlesworth (2014), Sinclair (2012), Smith (2011), and Tauber (2014) showcased post-disaster architectural projects and emphasized architects' roles in these. These authors portrayed successful architects, depicting the main elements of their proposals, and assessed their achievements. While doing so, they also highlighted the limitations faced by the new generations of architects when engaging in emergency response. In general, these authors agree that the current curricula of undergraduate architecture courses do not prepare professionals for working in the humanitarian sector. Alan Ricks and Michael Murphy, founders of the Mass Design Group NGO, stand for an architecture that promotes justice and dignity, in which every project is a design process that has a social mission and entails community engagement. These architects argued that when evaluating architecture only from an aesthetic or historical point of view, the impact that the building has had on the targeted community is lost. If the political and social implications of the process of designing are ignored, the recognition of great architecture is chocked (Murphy & Ricks, 2013).

Practitioners and scholars cited above seem to agree that architecture schools and professional boards still have to face the challenge of training architects to work in disaster and development settings. This research argues, therefore, for a better alignment with the HA tenets, to enable and expand students' and architects' skills so that they can embrace design as an authorial but also community-shared and socioculturally bounded process.

3. Some regular design competitions tackling risk and resilience

This section covers the contribution of design competitions to making the connection between the design, disaster, and development domains. As disaster risk, resilience, and climate change are nowadays increasingly omnipresent themes, these terms have become emergent keywords in contemporary design competitions. This fact actually configures an opportunity to bring the uncertainty and complexity inherent to these topics to a professional domain strongly marked by the search for certainties (Hobeica & Hobeica, 2019; Till, 2009). Many of these design competitions have been triggered by a focusing disaster event that gave prominence and a sense of urgency to the management of the related risks. Consequently, these competitions often have only a single edition and target the concept stage, being taken as architectural references for future interventions.

Some well-known examples of HA and DRR-related design competitions include Rebuild by Design, held between 2013 and 2014, focusing on Greater New York after the landfall of Hurricane Sandy (Ovink, 2016), and the 2018 Resilient Homes Challenge, jointly organized by the World Bank and the Build Academy, with an international scope and three hypothetical risk scenarios. Mostly targeted at architecture students, but also welcoming urban-planning, civil-engineering, landscape-design, and disaster-management multidisciplinary teams, i-Rec and DRIA are among the few HA and DRR-related competitions that are recurrent. In the following subsections, we examine the last five editions of each of these competitions, covering both their rationales and major outcomes.

3.1 i-Rec's experience (2010−2019)

Originated in the IF Research Group of the University of Montreal (Canada), i-Rec is an international network of multidisciplinary experts who work with post-disaster reconstruction and DRR. This network links academic, international, and community-based organizations with the aim of "contributing to the betterment of living conditions in human settlements," with a special focus on housing issues in the Global South (i-Rec, 2008). Since its inception in 2002, i-Rec has been organizing an international conference biannually, in which its members share experiences and give voice to affected vulnerable communities worldwide. The conference runs as a side event an architecture competition, as a means to promote related education and training. The i-Rec Student Competition is open to students from any university and offers prize money for the best projects. Having reached its ninth edition in 2019, the i-Rec competition is considered by the network's members as "an important reference for architectural education in the area of human settlements in developing countries" (i-Rec, 2008).

Regarding the design brief's features, i-Rec proposes an open scenario within a post-disaster setting. The proposals should necessarily consider the communities' participation and also care for the reconstruction process as a whole, including logistics and financing issues, besides architectural design. This competition brings some sort of realism to the architectural training, by introducing contingencies (the disaster context) and by requiring pragmatism (the actual means to make the solutions happen) from the submissions.

In the i-Rec competition's first edition in 2002, students from a variety of disciplines—from architecture and industrial design to urban planning and engineering—were challenged to propose housing strategies for any hazard-prone developing country. As evidenced by the profile of the last five editions, the core characteristics of the competition have been evolving. For instance, the enlargement of the scope of some features is noticeable, such as the work languages of the submissions (five languages besides English were accepted in the last edition), the geographical focus (although still targeting developing contexts, post-disaster proposals in developed countries are also welcome), and the acceptance of DRR-dedicated proposals not necessarily included in reconstruction projects. In contrast, just like the first edition, each subsequent one has specialized in a thematic focus (such as urban settings or displacement) and has thus proposed some particular design scenarios, which are nonetheless still open in terms of location and involved risks. In addition, the target of the competition has been most of the time restricted to students coming from architecture courses.

Out of the 12 winning entries of the last five years, eight focused on developing-country contexts, while five originated in a university based in such countries. Regarding the projects' assessment, all the i-Rec editions adopted the same criteria: "1. Context; 2. Technological approach to buildings, infrastructure, and settlement from a perspective of resisting disaster risk creation; 3. Organizational approach and logistics" (i-Rec, 2019). An analysis of the jury reports showed that the awarded proposals gave the best responses in terms of environmental and cultural sensitivity, socio-spatial dimensions, vulnerability reduction, organizational structure of project management, community involvement, and incremental growth. In sum, the focus of the winning proposals extended beyond the classic Vitruvian triad (firmness, commodity, and delight) (Wotton, 1624/2010), to embrace a human-centered approach that acknowledges DRR as a genuine design input.

3.2 DRIA's experience (2015−2019)

Initiated in 2014 by the School of Design and Environment of the National University of Singapore, DRIA is an international research program that focuses on design education and capacity building toward the increase of resilience in Asian cities, considering their high exposure to hazards (namely hydrological ones and earthquakes). Presently gathering 21 universities (two-thirds of which in Asia), DRIA seeks to support tackling climate change and the ensuing risks, through both mitigation and adaptation (DRIA, 2019a, 2019b). Since 2015, DRIA has been organizing an annual international symposium, in which invited international experts share their knowledge with members of the consortium's partners. This event also comprises a design competition, limited to a maximum of 20 proposals, open only to students enrolled in one of the following courses: urban design, architecture, urban planning, landscape architecture, and building science, in one of the partner universities. Backed by strong institutional sponsorships, the DRIA competition offers prize money and a trophy in three categories. The competition also covers the attendance of representatives of the participating teams in the conference.

In each edition, the DRIA competition proposes a single site to all design teams, targeting different Asian contexts, ranging from small villages (such as Xinxing Harbor Village in Hainan Province, China) to fragments of metropolises (such as the riverfront of Bangkok, Thailand). Urban and architectural design, together with building technologies, is called upon to ensure and enhance communities' resilience before hazards strike (pre-disaster scenarios) and also during disaster events. Since its second edition, the DRIA competition presents a detailed design brief, which includes a large set of information (such as maps, plans, and demographic data) related to the target setting. Since 2017 a motto has been associated with each edition: "Cities sinking in drought," "Drowning by the sea," and "Growing cities, shrinking waters" (DRIA, 2019b). Up to now, all the targeted design contexts were markedly water-related landscapes, thus often exposed to floods. Cyclone risk has also been a recurrent topic.

An analysis of DRIA's five design briefs reveals the increasing breadth of both objectives and the design scope, involving areas with complex risk profiles that require a multi-hazard strategy. Moreover, given the scale of the concerned areas, the students' proposals necessarily have to blur the boundaries between architectural, landscape, and urban design. Indeed, the adoption of a multiscale approach is among the evaluation criteria, which also include the understanding of the involved urban and climate-change challenges, as well as the quality of the approach and design solutions to address these challenges. The latter criterion comprises the comprehensiveness, innovativeness, and inclusiveness of the proposed features, and how they support adaptation, mitigation, and the reinstatement of environmental qualities (DRIA, 2019a). The proposed projects should also care for replicability, scalability, and for allowing measuring and quantifying their impacts. Furthermore, the projects should be technologically sound, as regards the adequacy of construction materials and systems, blending local traditions and new ideas.

According to the analysis made by the architect Poerbo (2018), who attended the first three editions of the DRIA symposium, the most common resilience strategies proposed in the 55 projects submitted to the competition up to 2017 were as follows: (1) the use of particular housing solutions (floating, cantilever, stilt, vertical, modular, or flexible); (2) the resort to raised platforms for the creation of public places; (3) the inclusion of community shelters in the proposed plans; and (4) the provision of community spaces and marketplaces. At the landscape scale, reforestation and other water-related interventions (such as drainage solutions or the creation of waterways) were among the more frequently proposed strategies (Poerbo, 2018). Such diversity reflects not only the design briefs' requirements but also the approach actually needed to comprehensively foster resilience.

Overall, it is interesting to note that most design teams who participated in the i-Rec and DRIA competitions come from ordinary architecture courses (that is, undergraduate level with a general scope), thus allowing one to infer that DRR is somehow latently present in the design curricula of the related schools. Moreover, in the case of DRIA, having the same schools participating in the competition potentially fosters the creation of a resilient-design culture among them. The model followed by DRIA of having the same site for all competitors provides better means for comparing the proposals, yet the teams are likely to design more contextualized solutions when they choose the setting themselves (the i-Rec model). Finally, although DRIA's rationale is centered on communities' resilience, an emphasis on technological means may hinder a more socially engaged design approach. Conversely, the i-Rec competition more clearly acknowledges the importance of the social dimension, giving a central role to design as a process with communities' participation.

4. The Building 4Humanity Design Competition

Drawing upon the experiences of i-Rec and DRIA, the Building 4Humanity Design Competition (B4H-DC) was launched by the Building 4Humanity NGO, as a side event of the eighth edition of the International Conference on Building Resilience (ICBR), held in Lisbon in November 2018. The B4H-DC thus benefited from the network of DRR professionals who had participated in previous editions of the ICBR.

The B4H-DC targeted interdisciplinary professional and student design teams. Although open to diverse professionals, the competition brief stipulated that each student team had to be supervised by a qualified architect. Just like in the i-Rec case, the teams had to present their own design programs, identifying a disaster-prone area and proposing alternatives on how adapted buildings and settlements could support the strengthening of communities' resilience. The projects could focus on either building better from the start or building back better, thus embracing risk mitigation and the spirit of both the Sendai Framework for Disaster Risk Reduction (UNISDR, 2015) and the 2030 Agenda's Sustainable Development Goal 11—"making cities and human settlements inclusive, safe, resilient and sustainable" (UN, 2015).

The competition had three categories open to any hazard type: professional—built and unbuilt—projects, and student projects. It received 60 entries from 33 countries spread over the 5 continents. Each proposal was remotely evaluated by at least 5 members of an international jury composed of 55 academics and practitioners from different professional fields. More than half of the jurors were DRR specialists participating in the scientific committee of the 8th ICBR. The shortlisted proposals went through a second and final assessment by a 12-member jury, during the conference in Lisbon, using the same criteria adopted in the first evaluation stage:

- the quality of design, with a particular emphasis on the robustness of the preliminary risk assessment and the ensuing resilience strategies and solutions proposed;
- the quality of the overall framing and bonding of the design solution within the site/settlement, considering the utilization of local resources;
- the evidence and adequacy of design strategies for the engagement of concerned communities; and
- the value of the construction features, namely feasibility, cost-effectiveness, sustainability, and adaptability.

Despite having only three categories, the evaluation of the proposals posed some challenges, namely related to the adoption of the same criteria for all entries, irrespective of having both architectural and urban-design projects competing within a category. Indeed, single buildings and territories have very different risk and resilience constraints and opportunities. In the following sections, based on the projects' plans and reports, we present the major lessons learned with the winning entries of the student category in both editions of the B4H-DC.

4.1 Winning entries of the first edition of the B4H-DC (2018)

The three winning student entries of the first B4H-DC depicted very different risk scenarios. Table 6.1 presents a summary of the characteristics of these teams and their proposals. In the case of the first-prize winning project, it is noteworthy that although it was originally elaborated within an academic design studio, this urban-design output was partially executed by the students, teachers, and community members, who raised the needed funds through crowdsourcing. In its turn, the runner-up project was one of the six proposals submitted by the same Guatemalan University, five of which tackling volcano risk and resettlement. The eruption of the Fuego Volcano in Guatemala in June 2018 functioned in fact as a trigger for the adoption of such a DRR-related program. Finally, the third-place winning proposal managed to handle two HA challenges at once: the resettlement of Rohingya refugees and an insular site highly prone to disasters.

Difficult access—steep and narrow pathways without proper pavement, if any—, private appropriation of collective spaces, and children's use are among the major challenges tackled in the Brazilian project. Yet, the proposal intended not only to provide a public-space design but also to create an alternative methodology for intervening in favelas' straight alleys (*becos*), understood as the "democratization of architecture through self-managed actions to

TABLE 6.1 Summary of the winning design entries of the first B4H-DC—student category.

	First prize	Second prize	Third prize
Team members	Pérola Barbosa, Ana Dresler, Raquel Penna, and Patrícia Monteiro Santoro dos Santos	Lily Chen, Valerio Lee, Alejandra Lima, Jorge Pérez, and Gerardo Rodas	Vishal Kumar, Akhilesh Shisodia, Mohit Arya, and Reva Saksena
Advisor	Pablo Benetti	Eduardo Andrade Abularach	Sanjeev Singh
University, city and country	Universidade Federal do Rio de Janeiro, Rio de Janeiro, Brazil	Universidad Rafael Landívar, Guatemala city, Guatemala	School of Planning and Architecture, Bhopal, India
Focused area	Morro do Alemão, Rio de Janeiro, Brazil	Finca La Industria, Guatemala	Thengar Char Island, Bangladesh
Focused risks	Landslides	Volcano eruptions	Cyclones and floods
Design program	Improving accessibility and livability of alleys in a consolidated slum	Resettlement of a community affected by the eruption of the Fuego Volcano	Resettlement of Rohingya refugees in a flood- and cyclone-prone island

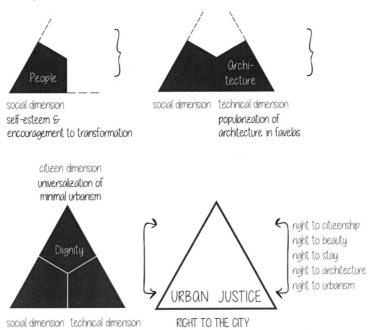

Design with a goal

ressigniy and transform the relationship of the local community with the Beco (the alley), through the design of the micro-scale and punctual interventions in the space

FIGURE 6.1   The conceptual framework adopted by the winning proposal. *Credit: Courtesy of the proponent team.*

empower the residents" (Fig. 6.1). As reported by the team, the underlying idea was that several micro-interventions in the public realm, spread all over an area, "may re-signify the territory in its entirety."

The "Seed Project" by the Guatemalan team took an interesting standpoint regarding time, acknowledging different post-disaster demands. Their urban plan contained activities to run within three schedules, corresponding to the emergency, transition, and development stages: "48 hours / 48 weeks / 48 months" (Fig. 6.2). These activities were finally associated with the six themes identified as priorities by the involved community, during a real participatory exercise. According to the team, the integration of the community's opinions and aspirations is needed "to secure a resilient development" after the eruption. Their proposal further developed the housing typology, in which row houses provide safety and spatial quality, and concomitantly promote social bonds.

The Indian team focused on the particularities of the Rohingya people, rendered "stateless and obscure in dilapidated camps, entrapped in a state of temporary permanence," and the Bangladeshi alternative to settling Rohingya refugees on an isolated island. To manage such a complex scenario, the project sought to articulate the demands of both resilience and

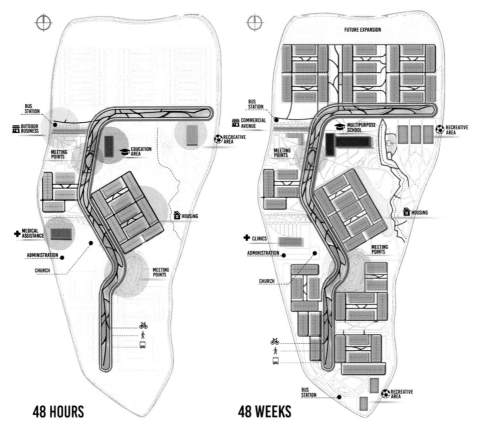

FIGURE 6.2 Two spatial expressions of the "48 hours / 48 weeks / 48 months" plan. *Credit: Courtesy of the proponent team.*

sustainable development to propose safe and friendly living spaces (Fig. 6.3), in opposition to the often-hostile environments encountered by involuntarily displaced populations. Their project thus handled three scales at once: the settlement, the cluster and the private dwelling—the latter based on the memories from Rohingyas' previous houses.

Despite their differences, these three projects exhibited remarkable contact points. For instance, the proposals for Bangladesh and Guatemala adopted a landscape perspective, addressing a large area surrounding the chosen design site. At the same time, they embraced an interplay between design scales, in which the housing proposal is strongly linked to the layout of the settlement. Conversely, the intervention in the Morro do Alemão favela employed a microscale lens but also considered the cumulative impacts of its approach at a larger scale. All three proposals derived from a detailed analysis of the involved site and sociocultural fabric. They implicitly absorbed the idea of design as a source of spatial seeds that can flourish to fulfill deep human needs. For example, Fig. 6.4 illustrates the transformation of a rather desolate and unsafe space into an attractive one, resulting from an actual intervention carried out by the Brazilian team.

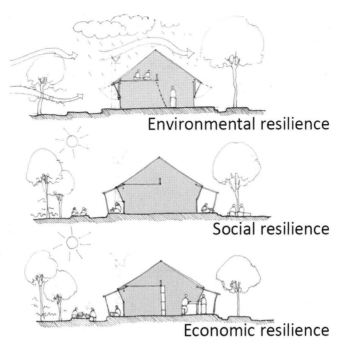

FIGURE 6.3 Fostering resilience in the three sustainable-development dimensions. *Credit: Courtesy of the proponent team.*

FIGURE 6.4 Colors, greenery, and art add to the livability in a *beco* of the Morro do Alemão favela. *Credit: Courtesy of the proponent team.*

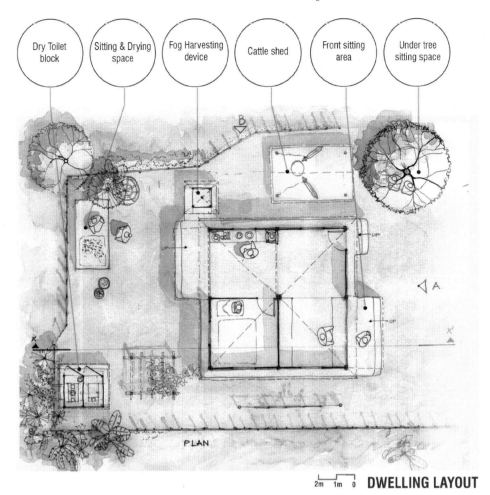

FIGURE 6.5   Simplicity and self-reliance as sustainable-living means in the Indian proposal. *Credit: Courtesy of the proponent team.*

In line with the design brief, the recurrent use of local materials and building technologies is also noteworthy. Moreover, the three proposals revealed a comprehensive approach to sustainability. Besides sustainable construction and environment-friendly urban design (Fig. 6.5), the projects also addressed social resilience through the active participation of local residents, experienced in real terms by all teams. Accordingly, a sense of social justice pervades the winning projects. For instance, the three proposals expressed the search for equity and inclusion through their realism in terms of cost-effectiveness and construction phasing. The possibility of incremental expansion of the houses was also a key feature addressed by the Indian and Guatemalan teams (Fig. 6.6).

Beyond their particularities in terms of geographies, landscapes, and DRR demands, these projects' approaches to HA are recognizable in their common willingness to match the

FIGURE 6.6 Housing proposal to safely live at the bottom of the Fuego Volcano. *Credit: Courtesy of the proponent team.*

aspirations of local people. It does not matter whether these are refugees, disaster-affected communities, or urban dwellers exposed to violence and unhealthy conditions. Whatever the addressed issues, these projects envisaged the creation of living spaces that enable the reinforcement of social, cultural, and symbolic bonds within the targeted communities. They prioritized equity and empowerment, regardless of the aesthetic output, thus highlighting the power of good design as a tool to benefit the most vulnerable.

4.2 Winning entries of the second edition of the B4H-DC (2019)

Jointly organized by B4H and the Taiwanese NGO Architecture for People, the second B4H-DC followed an approach different from the previous edition's one. Similarly to DRIA, the organizers selected a 0.7-ha site and invited the participants to respond to a particular program. The design should focus on the sheltering of 400 Syrian refugees and their inclusion in the Turkish host city of Reyhanli, near the Syrian border. Expanding the framework of the 2018 edition, the new design brief incorporated specific issues related to forcedly displaced persons and climate change adaptation. Regarding the composition of the jury, the 2019 edition featured only eight personalities—experienced and awarded scholars and practitioners—, but only two of them had strong links with HA. Finally, the jurors met and worked during a week to assess the 140 applications—80 of which belonging to the student category.

Maybe as a result of the high profile of the jury composition, which comprised the prominent Finnish architect and theoretician, Juhani Pallasmaa (former juror of the Pritzker Prize) and Suha Ozkan (former secretary-general of the Aga Khan Award for Architecture) the three winning projects, somehow exalted design as a distinctiveness mark. Yet, two of these projects also highlighted a particular sense of community. Although not tackling in an entirely convincing manner the participation of the targeted refugee community, the local residents, or other stakeholders, the projects depict a sense of place through the design of public spaces in a good balance with the shelter clusters. These projects relied on the use of a polyhedral structural framework in which the provisional shelters could be encapsulated, replicated, and scaled up. The smart use of the incremental-housing concept, key in HA projects, seemed to have resonated in the jury's appraisal. On the other hand, the runner-up project, as well as one of the honorable-mention proposals (Fig. 6.7), stood for strong architectural gestures, embodying values that impressed the jurors, such as freedom, nature, and mobility.

The first and third winning projects have tangible and intangible similarities. Modular structures ensure a grid system that accommodates different closed, open, and empty spaces for various uses while anticipating an easy assemblage with local recycled materials (such as wood sandwich panels filled with sheep wool and palettes) (Fig. 6.8). The two projects

FIGURE 6.7   The honorable-mention proposal by Constanza Soto, María Paz Farías, Carolina Alejandra Carvajal, Javiera José Jaramillo, and Stephanie Gabriela Reveco, from Universidad de Valparaíso (Chile). *Credit: Courtesy of the proponent team.*

FIGURE 6.8   The third-prize proposal, by Çağan Köksal, Barış Açık, Burak Arifoğlu, Begüm Yoldas, and Akın Ertürk, from Kadir Has University (Turkey). *Credit: Courtesy of the proponent team.*

FIGURE 6.9   The first-prize proposal, by Elif Arpa, from the Department of Sociology of Boğaziçi University (Turkey). *Credit: Courtesy of the proponent team.*

FIGURE 6.10   The second-prize proposal, by Ariel Gajardo Barahona, Pablo Cantillana, Osvaldo Garrido Parada, Alejandro Olives, and Danilo Reyes, from Universidad de Valparaíso (Chile). *Credit: Courtesy of the proponent team.*

reserved a significant role for refugees in the assemblage and management of water-collection and energy systems. Both projects depicted a full range of solutions related to lifestyle— identifying spaces for keeping traditional activities such as drying fruits—, sharing— encouraging the exchange of goods through indoor and outdoor facilities such as urban furniture—, and to livelihoods—allowing for spaces suitable for marketing and businesses inside the shelter clusters (Fig. 6.9). In a different register, the runner-up project suggested a powerful link to refugees' collective memory through a sharp contrast between a temporary textile cover and a permanent built floor (Fig. 6.10). The latter was intended to last beyond the shelters and become the ground for a future urban setting, therefore being useful for the local population.

The awarded projects presented singular features related to inclusion and integration that go beyond the brief's requirements. They put forward new ideas on how to house refugees with dignity, meeting the needs of minority and vulnerable groups, such as the wounded, the young, and the elder. The projects still provided private and public spaces to accommodate social and economic activities that fit the Syrian culture. However, this match between architecture and the refugees' social resilience raised many questions on the fulfillment of DRR requirements. Perhaps given the relatively low level of concerns in this area regarding natural hazards, when compared with other Turkish regions, the projects posited very few solutions for disaster preparedness and DRR. In this sense, they eventually did not entirely meet the competition brief's requirements, which explicitly recommended observing the Sendai Framework.

5. Discussion

Given the contrasting focuses of the two editions of the B4H-DC, it would not be feasible to compare their winning proposals. Yet, it is possible to analyze them in an aggregated manner to assess their contributions to enhance the HA practice and also to take lessons for architectural education. Firstly, the outcomes of both editions demonstrated the students' ability to address HA issues, be them related to emergency or development contexts. Just like in the i-Rec and DRIA competitions, none of the B4H-DC's winning teams came from specialized master programs related to disaster, design, and development. Thus, their HA abilities constitute empirical evidence that such design programs should be acknowledged and stimulated at architecture schools.

Secondly, the winning proposals in both editions of the B4H-DC demonstrated high levels of originality regarding architectural solutions to boost community engagement and resilience. Indeed, these people-oriented projects somehow exemplify a choice for design with empathy (see Chapter 8, by Sandman and Suomela, in this volume). Although design projects can never tackle all contextual uncertainties, these proposals pinpoint that architects can develop higher levels of empathy and "design like they really give a damn" (Sinclair, 2012). In the case of the 2018 edition, it is notable that the projects did not emphasize design as a mere aesthetic concern but weighed the brief's requirements to stress DRR features and prioritize community resilience. These two concerns clearly diminished the eye-catching character of the proposals but offered an opportunity for more articulated designs that harmonically blended aesthetic, sociocultural, environmental, and economic aspects.

Our analyses also indicate the value of involving students in the formulation of the architectural program. Whether they are encouraged to draw up the architectural problem (such as in i-Rec or the 2018 edition of the B4H-DC), or they have the opportunity to define the target groups of their proposals (such as in the 2019 edition of the B4H-DC), their projects tend to be focused and complex, reaching a high-quality level. Thus, there seems to be room for reinforcing creativity by detaching from conventional architectural programs and promoting cutting-edge interdisciplinary exercises that reflect some of the most pressing global societal challenges of our times.

The experiences of i-Rec, DRIA, and B4H-DC put in evidence that the design brief and interdisciplinarity are key ingredients of competitions addressing HA issues. When the design brief is aligned with the requirements of a comprehensive HA approach, the participating students can develop their best capacities and spur their imagination. A design brief that demands a good balance between technical, environmental, social, economic, and community-participation aspects, while providing the freedom to wander morphologically and aesthetically, may prompt powerful and creative ideas. Such a balance requires interdisciplinarity as a prerequisite for good HA projects. When featuring a compelling brief and experienced jurors, and attracting skillful and varied team members, design competitions can become an irreplaceable opportunity to unleash the best ideas from the participating students. In this sense, HA can constitute a fertile field to be explored with conceptual experimentations.

6. Conclusions

Many of the authors quoted in this chapter broadly claimed that the lack of preparation of architects in the humanitarian arena stemmed from the inadequacy of the education received in architecture schools. Some of these scholars have been willing to change this unsuitable state and support moving toward a sort of architecture training for engagement. The analyses in this chapter revealed that projects elaborated within HA and DRR-related competitions concomitantly cared for the spatial, social, and political dimensions of disaster-prone settings within the design task. Moreover, they interpreted stakeholders' capacities to support building resilience in an inclusive way. Thus, they did not dissociate spaces' physical, sociocultural, and symbolic wholeness. Further, design competitions addressing HA can sensibly engage students while fostering both their problem-solving and design capacities. Indeed, these competitions call for architecture students to come closer to the role of atmosphere creators while meeting people's expectations.

References

Acar, E., & Yalçınkaya, F. (2016, July). Integrating disaster management perspective into architectural design education at undergraduate level: A case example from Turkey. In *Paper presented at the 5th World Construction Symposium: Greening Environment, Eco-innovations & Entrepreneurship, Colombo, Sri Lanka*. Retrieved from https://www.researchgate.net/publication/303988576_Integrating_disaster_management_perspective_into_architectural_design_education_at_undergraduate_level_-_A_case_example_from_Turkey.

Aquilino, M. (Ed.). (2011). *Beyond shelter: Architecture for crisis*. London: Thames and Hudson.

Bachelard, G. (1957/1994). *The poetics of space*. Boston, MA: Beacon Press.

Benetti, P., & Carvalho, S. (Orgs.) (2017). *Praça pr'Alemão ter: O germinar de uma praça verde no Morro do Alemão [A first-class square: The inception of a green square in Morro do Alemão]*. Rio de Janeiro: UFRJ-PROURB-FAU/Instituto Raízes em Movimento (in Portuguese).

Boano, C., & García, M. (2011). Lost in translation? The challenges of an equitable post-disaster reconstruction process: Lessons from Chile. *Environmental Hazards, 10*(3−4), 293−309. https://doi.org/10.1080/17477891.2011.594493.

Boano, C., & Hunter, W. (2012). Architecture at risk (?): The ambivalent nature of post-disaster practice. *Architectonica, 1*(1), 1−13. https://doi.org/10.1108/IJDRBE-07-2013-0025.

Brogden, L. (2019). Sustainability, design futuring, and the process of shelter and settlements. In A. Asgary (Ed.), *Resettlement challenges for displaced populations and refugees* (pp. 1−14). Cham, Switzerland: Springer. https://doi.org/10.1007/978-3-319-92498-4_1.

Charlesworth, E. (2014). *Humanitarian architecture: 15 stories of architects working after disaster*. Oxford: Routledge.

Charlesworth, E. (2015). Humanitarian architecture: Seeking spatial solutions for complex global challenges. In J. Mackee, H. Giggins, T. Gajendran, & S. Herron (Eds.), *Proceedings of the 5th International Conference on Building Resilience held in Newcastle City Hall, Newcastle, New South Wales, Australia*. Retrieved from https://www.newcastle.edu.au/__data/assets/pdf_file/0008/202967/FInal-5th-BRC-Proceedings-23-07-15.pdf.

Davis, I., & Alexander, D. (2016). *Recovery from disaster*. Abingdon, Oxon: Routledge.

DRIA. (2019a). *Designing Resilience in Asia International Design Competition: Growing Cities, Shrinking Waters [Design brief]*. Retrieved from https://drive.google.com/file/d/1bBBJrkG1TyoXhVg-P4yuCHdFqCVUepCg/view.

DRIA. (2019b). *DRIA 2019 NCKU [Book]*. Retrieved from https://www.dropbox.com/s/cwsr6blcaghl2fe/2019 DRIABOOK2019.pdf.

Durand, J.-N.-L., & Moneo, R. (1981). *Compendio de lecciones de arquitectura. [Summary of architecture lessons.]*. Madrid: PRONAOS (in Spanish). (Original work published in French in 1809), as *Précis des leçons d'architecture données à l'École Royale Polytechnique*).

Ehrmann, S. (Ed.). (2018). *Design, Disaster & Development Research Forum: How do we deal with the pedagogic, spatial and research challenges of global mobility, migration and social inequality?* (Report), Retrieved from http://masteremergencyarchitecture.com/2018/10/08/key-lessons-from-rmits-design-disaster-and-development-forum.

Hamdi, N. (2004). *Small change: About the art of practice and the limits of planning in cities*. New York, NY: Routledge.

Harris, V. (2011). The architecture of risk. In M. Aquilino (Ed.), *Beyond shelter: Architecture for crisis* (pp. 12–24). London: Thames and Hudson.

Hobeica, L., & Hobeica, A. (2019). How adapted are built-environment professionals to flood adaptation? *International Journal of Disaster Resilience in the Built Environment, 10*(4), 248–259. https://doi.org/10.1108/IJDRBE-06-2019-0029.

i-Rec. (2008). *i-Rec program*. Retrieved from http://www.grif.umontreal.ca/pages/i-Rec-program.pdf.

i-Rec. (2019). *9th Student Competition — Iatrogenesis*. Disrupting the status quo: Resisting disaster risk creation (Jury report). Retrieved from http://membresirec.umontreal.ca/student_competition/9th/20190610_CompetitionReport.pdf.

Lepik, A. (2010). *Small scale, big change: New architectures of social engagement*. New York, NY: MoMA.

Lizarralde, G., Johnson, C., & Davidson, C. (Eds.). (2010). *Rebuilding after disasters: From emergency to sustainability*. Oxford: Spon Press.

Martins, A. N., & Guedes, M. C. (2015). 'Humanitarian' or 'resilient architecture' for vulnerable communities? In J. Mackee, H. Giggins, T. Gajendran, & S. Herron (Eds.), *Proceedings of the 5th International Conference on Building Resilience in Newcastle City Hall, Newcastle, New South Wales, Australia*. Retrieved from https://www.newcastle.edu.au/__data/assets/pdf_file/0008/202967/FInal-5th-BRC-Proceedings-23-07-15.pdf.

Martins, A. N., & Rocha, A. (2019). Risk and resilient architectural practices in informal settlements: The role of NGOs. *International Journal of Disaster Resilience in the Built Environment, 10*(4), 276–288. https://doi.org/10.1108/IJDRBE-09-2019-0063.

Muratori, S. (1960). *Studi per una operante storia urbana di Venezia* [Studies for an active urban history of Venice]. Rome: Istituto Poligrafico dello Stato (in Italian).

Murphy, M. P., Jr., & Ricks, A. (2013). Review of the book *Beyond shelter: Architecture and human dignity*. by M. Aquilino (Ed.). *The Journal of Architecture, 18*(1), 111–114. https://doi.org/10.1080/13602365.2013.767054.

Narayanan, N. P. (2013). Design as understanding: Illustrations from an academic experiment. In D. Ramirez-Lovering, J. Alexander, & A. Fairley (Eds.), *Proceedings of the 7th International Conference of the Association of Architecture Schools of Australasia* (pp. 62–75). Melbourne, Australia: The Association of Architecture Schools of Australasia. https://doi.org/10.13140/RG.2.1.4460.0805.

Norberg-Schulz, C. (1980). *Genius loci: Towards a phenomenology of architecture*. New York, NY: Rizzoli.

Ovink, H. (2016). Redesigning the design competition. *Urban Solutions, 9*, 42–51.

Pallasmaa, J. (2014). Space, place and atmosphere: Emotion and peripherical perception in architectural experience. *Lebenswelt, 4*, 230–245. https://doi.org/10.13130/2240-9599/4202.

Pallasmaa, J. (2018). *Juhani Pallasmaa interview: Art and architecture [Video file]*. Retrieved from https://vimeo.com/270345281.

Poerbo, H. W. (2018). Recurring design concepts for resiliency in Asia. *IOP Conference Series: Earth and Environmental Science, 152*(1). https://doi.org/10.1088/1755-1315/152/1/012003.

Sanderson, D. (2010, March 3). Architects are often the last people needed in disaster reconstruction. *The Guardian*. Retrieved from http://www.theguardian.com.

Sinclair, C. (2012). Introduction: Lesson learned. In *Architecture for Humanity, Design like you give a damn: Architectural responses to humanitarian crises, 2* pp. 11–47). Metropolis Books, New York, NY: Abrams.

Smith, C. (Ed.). (2011). *Design with the other 90%: Cities*. New York, NY: Smithsonian Cooper-Hewitt, National Design Museum.

Tauber, G. (2014). *Architects and post-disaster housing: A comparative study in South India*. Bielefeld: Transcript.

Till, J. (2009). *Architecture depends*. London: MIT.

Tovivich, S. (2010). *Architecture for the urban poor, the 'new professionalism' of 'community architects' and the implications for architectural education: Reflections on practice from Thailand* (Doctoral thesis). London, United Kingdom: University College London. Retrieved from https://discovery.ucl.ac.uk/id/eprint/1306880/1/1306880.pdf.

UN (United Nations). (2015). *Transforming our world: The 2030 Agenda for Sustainable Development.* Retrieved from https://www.un.org/ga/search/view_doc.asp?symbol=A/RES/70/1&Lang=E.

UNISDR (United Nations Office for Disaster Risk Reduction). (2015). *Sendai Framework for Disaster Risk Reduction 2015–2030.* Retrieved from https://www.preventionweb.net/files/43291_sendaiframeworkfordrren.pdf.

Wagemann, E., & Ramage, M. (2013). Relief for the curriculum: Architecture education and disaster recovery. *Scroope: Cambridge Architecture Journal, 22*, 129–133. https://doi.org/10.17863/CAM.6437.

Wotton, H. (1624/2010). *The elements of architecture, collected by Henry Wotton Knight, from the best authors and examples.* London: John Bill (Reprinted in 2010).

Zumthor, P. (2006). *Atmospheres: Architectural environments, surrounding objects* (5th ed.). Basel: Birkhäuser.

Architecture and urban design to enhance community preparedness

7

Architects' multifaceted roles in enhancing resilience after disasters

Anouck Andriessen, Angeliki Paidakaki, Cynthia Susilo, Pieter van den Broeck

KU Leuven, Leuven, Belgium

O U T L I N E

1. Introduction

A wide-ranging discussion within the disaster scholarship and practice focuses on the social, economic, and cultural repercussions of reconstruction strategies, such as economic and social impoverishment, marginalization, and disruption of social networks. These failures are often the result of insufficient incorporation of social and cultural specificities into the design of recovery strategies and programs, which inevitably affects the vulnerability of the concerned communities. Several authors have claimed architects' important role in addressing these negative repercussions when the design is performed as a transformative instrument and process in which empowerment, agency, and alternatives are central ideas in the rebuilding of post-disaster communities (Boano & Garcia, 2011; Boano & Hunter, 2012; Luansang, Boonmahathanakorn, & Domingo-Price, 2012). Nevertheless, the multifaceted role of architects and their position in the complex web of stakeholders in post-disaster resilience processes have been insufficiently deconstructed and discussed in the disaster discourse.

To address this literature lacuna, in this chapter, we analyze the various roles architects play in fostering the resilience of disaster-affected houses and communities in and through diversely governed post-disaster reconstruction projects. More specifically, we seek answers to the following questions: What are the varied roles architects play in resilience-building processes? To what extent does the governance structure of a recovery project foster or hinder the potential contribution of architects to building resilience? How can architects overcome institutional rigidities to bolster and optimize the potential of their profession to building resilience in disaster-recovery contexts?

To answer these questions, we first conceptualize the resilience—architect nexus and deconstruct the multifaceted manner in which architects enhance community resilience by gaining essential inspiration from the community-architecture scholarship (Aquilino, 2011; Charlesworth, 2014; Luansang et al., 2012; Tauber, 2015). In turn, we empirically examine how the resilience—architect nexus is unveiled in real-life recovery contexts by focusing on the implementation of three diversely governed reconstruction initiatives in the aftermath of the 2010 Merapi Volcano eruption in Indonesia. These initiatives are: a top-down program led by the international telecommunications company Qatar Telecom (Q-Tel); a bottom-linked program called REKOMPAK led by the state and financed by the World Bank; and a bottom-up program led by the community-architecture group ARKOM. The choice of these three programs was purposeful and dual: to disclose the wider possible panorama of the architects' roles in post-disaster resilience-building processes and to reveal the extent to which the governance modality of a recovery program shapes, bolsters, or limits the potential contribution of architects to building resilience.

In this empirical examination, resilience is evaluated on two scales: the community scale, entailing community-development processes, and the individual-housing scale, involving the process of making safe and affordable homes. These scales jointly clarify the full involvement of architects in reconstruction processes and the architects' impact on the lives of the beneficiaries. The same methodology is applied to all case studies, ensuring the uniformity of the results. The empirical investigation took place in Indonesia between February and March 2018 and consisted of the following: semistructured interviews with architects employed by various entities (for instance international organizations, the Indonesian Government, community-architecture organizations, and international companies), government officials (at the national, district, and local levels), engineers (working together with architects), local scholars, and elected village leaders; several field visits in the three recovery sites; and questionnaire-based interviews and informal conversations with community members. The collected data (transcribed interviews, notes from field observations, and other secondary materials) were organized in the form of a fieldwork report whereby the heterogeneity of architects' roles was sketched out. The most pertinent information to our research objective was later extracted from the report, and a rigorous content analysis was carried out.

This chapter is hereafter structured as follows. In Section 2, we delve into the literature of resilience and community architecture to clarify the notion of resilience from the standpoint of architects and develop an inceptive overview of the different roles architects play in post-disaster resilience-building processes. Section 3 confronts the theoretical investigation with the studied cases to uncover the roles architects played in the post-Merapi eruption reconstruction processes. In Section 4, we discuss how architects contributed to building the resilience of the affected communities and how the governance structure of each project fostered or hindered the effectiveness of this role. Finally, Section 5 provides some general conclusions and future research potentials.

2. Conceptualizing the multifaceted community-architecture practice

This section sets out with clarifying the practice of community architecture and focuses on how architects can contribute to resilience building at both the housing and community scales. To achieve this, we bring the literature of resilience into conversation with the currently existing literature on the involvement of architects in post-disaster contexts.

2.1 Community architecture

The architects involved in post-disaster projects are often referred to as "humanitarian" or "community" architects. Community architecture was first described as an unconventional architectural practice that includes end-users in the design and production of their own built environment (Habraken & Teicher, 1972; Turner, 1972, 1980; Wates & Knevitt, 1987). Currently, community architecture is commonly applied in the context of post-disaster reconstruction, using participatory design to build more resilient communities. Authors such as Charlesworth (2007, 2014), Aquilino (2011), Stohr and Sinclair (Architecture for Humanity, 2012), Boano et al. (Boano & García, 2011; Boano & Talocci, 2017) have, on the one hand, placed great emphasis on the merits of community architecture in the post-disaster sphere. These include participatory design, community mapping, bottom-up approaches, and collective building to nurture the existing strengths of disaster-affected communities (traditional building techniques, social networks) and contribute to creating safer, more democratic, and culturally relevant post-disaster living environments (Luansang et al., 2012). On the other hand, the main constraints of the practice have been identified, the most important of which are the limited societal impact of small-scale interventions and the lack of sufficient funding (Charlesworth, 2014; Coulombel, 2011).

2.2 Architecture for resilience building

Our interpretation of architecture for resilience originates from the definition of resilience as the capacity to "bounce forward" that underpins the evolutionary dimension of resilience (Manyena, O'Brien, O'Keefe, & Rose, 2011; Paidakaki & Moulaert, 2017). From this perspective, resilience is not only interpreted as a fixed capacity but also seen as a continually changing, socially transformative process (Davoudi, 2012), with various "bounce-forward" imaginations (Paidakaki & Moulaert, 2018). As Paidakaki and Moulaert (2017, p. 278) explained, this approach underscores the importance of tackling the underlying chronic social determinants of vulnerability as well as reflecting the long-term social learning, institutional adaptation, and social transformations triggered by a disaster and maintained across time, guiding future disaster-planning trajectories.

How can architects contribute to this resilience-building process, and promote and materialize novel recovery trajectories? Post-disaster reconstruction and community resilience building are very complex endeavors for which architects can use multiple professional skills and competences. The contribution of architects in building up the resilience of the affected population is dual. On the one hand, architects intervene physically in the built environment and design constructions that are strong and capable to withstand future hazards. On the other hand, architects support an open recovery process (Boano & Garcia, 2011), socially engaging with the affected communities and multilevel institutions and incorporating cultural and economic aspects of the disaster-affected communities in their designs.

2.2.1 Architecture for resilience building at the community level

Echoing the works of Oliver-Smith (1991) and Lorenz (2013), architects can contribute to building community resilience by considering the following factors:

- A careful choice of the site, in case relocation is preferred. The new site should be embedded in the existing urban fabric and in close distance from other family members;
- The layout of the village must fit into the needs and social-physical fabric of the affected community and integrate disaster preparedness to raise the community's awareness vis-à-vis future disasters;
- Economic variability must be integrated into the design through the provision of new livelihood opportunities as the diversification of income sources increases people's ability to maintain a steady income in crisis times;
- The relationships inside heterogeneous communities and between communities and other stakeholders (e.g., government, donor entities, and NGOs) must be strengthened to facilitate productive interactions and exchanges of knowledge and skills;
- Community participation should be fostered throughout the design process. The exact form of participation must reflect the political and sociocultural context of the disaster area; and
- The adaptive capacity of communities and their reliance on pre-disaster social networks must be respected.

Although architects are expected to take these factors on board, the feasibility of such an endeavor highly depends on multiple parameters such as the nature, interests, and visions of other stakeholders present in the recovery of the post-disaster area, and the availability of suitable land in relocation strategies. Time pressures and fund allocations, which are pertinent to disaster-reconstruction contexts, are also key conditions in terms of feasibility.

2.2.2 Architecture for resilience building at the housing level

When zooming in on resilience-building processes at the housing scale, architects can build resilience by taking the following factors into consideration:

— The structural resilience of physical constructions;
— A smooth transition from a post-disaster sense of "placelessness" to "home-making" and "belonging" (Zetter & Boano, 2009), by considering housing not solely as a material artifact but also as a process that has a significant impact on people's lives (Turner, 1972); and
— Community participation in the housing design whereby a two-way knowledge exchange is established: architects learn about traditional building techniques, local culture, and values, while the affected community learns about proper safe construction techniques from the architects (Jigyasu, 2008).

The exchange of knowledge and expertise between architects and local communities about appropriate housing reconstruction is, nevertheless, not always straightforward. Architects are often in favor of expert-led reconstruction and struggle to comprehend the value of traditional construction practices such as self-building. Additionally, local building techniques often get lost due to modernization and, in turn, are less trusted as safe by the affected population. Keeping a balance between tradition and modernization is a challenge for architects working in disaster situations.

2.3 "Mapping out" the resilience potential of the multifaceted community architect

Building upon this existing knowledge, in this section we outline an inceptive overview of the architects' various roles in post-disaster resilience-building processes. Each specific role is connected with its resilience potential at the community and the housing levels. This overview is presented in Table 7.1.

An important challenge for architects involved in reconstruction projects is not only to perform the aforementioned roles—individually or combined—to meet the specific needs and bolster the resilience of the affected communities, but also to test the feasibility of their practices within given sociocultural contexts and governance settings (Eizaguirre, Pradel, Terronesm, Martinez-Celorrio, & Garcia, 2012; Paidakaki & Moulaert, 2017; Pelling & Dill, 2010). Post-disaster recovery projects differ in nature and can be governed through top-down, bottom-linked, or bottom-up approaches. In top-down governance structures, the design ideas for reconstruction projects are imposed from the higher administration levels to the architects and communities at the recovery site. Conversely, bottom-linked projects allow platforms for interinstitutional exchanges across multilevel and multifarious recovery stakeholders—including architects—to foster a culture of participation in the recovery design, decision-making, and resource allocation (Eizaguirre et al., 2012). Lastly, bottom-up

TABLE 7.1   An overview of architects' various roles in post-disaster resilience-building processes.

Role of the architect	Description	Factors fostering or hindering resilience at two scales	
		Community level	Housing level
The genius designer	These professionals design autonomously, without considering the interdisciplinary character of the post-disaster context or the preexisting political conditions (Boano & Hunter, 2012). These architects seldom include community participation in the design. They, however, sometimes attempt to make a socially appropriate design through their own research initiatives	Since the community is not actively involved in the design process, the recovery outcome is oftentimes sociopolitically inappropriate because livelihood activities, stakeholder relations, and local dwelling culture are not integrated into post-disaster urban design	While the structural resilience of the physical structures can be adapted to the local building regulations, architects can neither incorporate the process of "home-making" (Zetter & Boano, 2009) into the design process nor foster a two-way knowledge exchange between themselves and the community
The building teacher	These architects strive for long-term sustainable solutions through knowledge transfers with the involved community in an easily understandable language. At the same time, these architects use locally available materials so that local people can easily purchase them in their future construction additions	By teaching the locals about construction techniques and using local materials, the economic activities of the community are promoted. The local dwelling culture is, however, not transferred to the architects to improve the urban layout	Long-term structural resilience is fostered because the community is taught to rebuild safely. In this role, nevertheless, the architect does not integrate the local specificities into the design
The attentive student	These professionals are aware of the fact that local communities have often designed their houses for generations (Luansang et al., 2012). As a result, they are open to learn from the local community through participatory processes that establish exchanges and incorporate such local knowledge into the design	The local dwelling culture is integrated into the urban layout, the knowledge of economic activities and the existing adaptive capacity are utilized to provide livelihood activities, and pre-disaster social relations of the community are accounted for	This one-way knowledge exchange fosters the process of "home-making" (Zetter & Boano, 2009) and integrates traditional building techniques, although these may be forgotten because of modernization
The compassionate friend	These architects wish to bridge the professional–people gap by cultivating a comfortable environment within the community. Once a bond is built between inhabitants and professionals, conflicts that exist in the heterogeneous	The knowledge of the power relations inside the community and their conflicts can ensure an equitable urban design, taking the vulnerability of people into account	Collaboration between architects and community members is enhanced when a bond is formed, fostering the process of "home-making" (Zetter & Boano, 2009) and learning from each other

TABLE 7.1   An overview of architects' various roles in post-disaster resilience-building processes.—cont'd

Role of the architect	Description	Factors fostering or hindering resilience at two scales	
		Community level	Housing level
The involved facilitator	community can be efficiently tackled throughout the design phase (Zetter & Boano, 2009) These professionals combine aspects of the attentive student and the building teacher to design together with the community, establishing a two-way knowledge exchange (community—professional and the other way round) through participatory processes. Considering the community as the most important actor in the post-disaster context, the involved facilitator helps to envision their possible futures	The location and layout of the site are codetermined by the community through participation processes, livelihood activities are nurtured, and pre-disaster adaptive capacities are identified and taken into consideration	Proper construction techniques within the local dwelling culture can lead to long-term sustainable, socioculturally adequate houses
The interfering mediator	These professionals negotiate between the affected community and other possibly involved stakeholders (e.g., government agencies, donors, NGOs). Using the reconstruction project to bring conflictive parties to the table, these architects serve as intermediaries in their effort to solve controversial views in tangible ways (Charlesworth, 2007)	Through multilevel negotiations, the affected community becomes more integrated into larger governance structures, increasing its involvement in the wider society	In this role, architects mostly focus on the community scale
The radical reformer	In this role, architects encourage the community to oppose current dominant powers, instigating changes toward a more equal society. Rather than including the excluded, they attempt to render current hegemonic powers inoperative (Boano & Talocci, 2017) and strive toward an urban reality based on equity. Radical reformers are politically involved and engage in critical urban issues (Charlesworth, 2007), striving for a reinvention of current architectural-design practices through their creativity (Aquilino, 2011)	The existing relations between the community and other stakeholders change through the architects' advocacy efforts toward decision- and policy-makers, which leads to the empowerment of the community	In this role, architects mostly focus on the community scale

(Continued)

TABLE 7.1 An overview of architects' various roles in post-disaster resilience-building processes.—cont'd

Role of the architect	Description	Factors fostering or hindering resilience at two scales	
		Community level	Housing level
The assisting architect	When the community is considered the architect of a reconstruction project, the roles of professionally trained architects are limited to providing trauma-healing projects (e.g., community mapping, cash-for-work initiatives) and technical assistance, empowering the community to strive for more inclusion into the wider society and creating networks in multilevel governance structures	This role attempts to incorporate all factors that foster resilience at this level	The house design by the inhabitants with the technical advice of building professionals fosters architecture for resilience at the housing level in all the previously mentioned factors: the structural resilience, the process of house-making, and a two-way knowledge exchange

initiatives are often initiated by organized citizens who seek an alliance with reconstruction stakeholders to put forward alternatives for a more equal and safer society (Eizaguirre et al., 2012).

The specific governance structure of a recovery project, we hypothesize, affects the roles and functions that resilience architects activate. In the following section, we make use of empirical evidence from Indonesia to examine this hypothesis and seek answers to our research questions.

3. The architects' role in post-Merapi eruption reconstruction programs

Our investigation of the multifaceted resilience architect in diversely governed recovery projects drew empirical evidence from three reconstruction projects in the Sleman Regency (Yogyakarta Province, Indonesia) after the 2010 Merapi eruption. Provinces in Indonesia are subdivided into regencies and cities, which are the key administrative units since the decentralization of the country in 2001. Regencies are further subdivided into districts and then again into villages. Mount Merapi is an active stratovolcano on the island of Java (see Fig. 7.1), located 30 km away from the city of Yogyakarta. Merapi is one of the most active volcanos in the world and erupted several times during a period of 10 days between October and November 2010. The largest of those eruptions caused the displacement of 350,000 people, the death of 368 people, severe damage in settlements and public infrastructure, as well as major interruptions in social and economic activities (Samekto & Nuh, 2017).

The housing stock was specifically affected, with 3023 damaged houses in the Sleman Regency from a total of 3931 damaged houses throughout the whole disaster area (Maly, Iuchi, & Nareswari, 2015). The reconstruction was mostly led by the government, which opted to relocate the entire affected population to safer areas and forbade any residency in the most

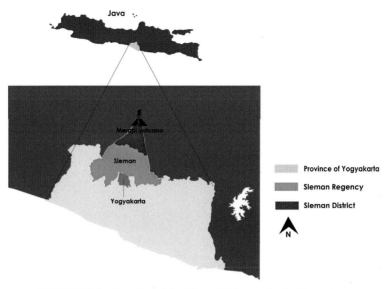

FIGURE 7.1 Location of the Merapi Volcano. *Credit: The authors.*

dangerous sites around the summit of the volcano (architect, personal communication, February 2018). Nevertheless, some of the affected communities, which viewed the volcano not as a threat but rather as a blessing that provides fertile grounds, refused to relocate.

3.1 Indonesian architects in disaster contexts

The involvement of Indonesian architects in disaster recovery is considerably new in the country as the post-disaster recovery of the 2010 Merapi eruption was only the third national-scale post-disaster recovery action involving them. Although the country is very prone to multiple environmental disasters, the knowledge relevant to disaster-responsive design remains lacking in local architectural and urban-design training, and in professional design practices.

Moreover, the architecture profession and practices in Indonesia generally tend to be trapped in their position as genius designers since the modernization manifesto of Arsitek Muda Indonesia (Young Indonesian Architects) in the mid-1990s. The manifesto has overly stressed the need of Indonesian architects to maintain their autonomous position in the society to be able to deliver independent and objective designs (Kusno, 2012). Although delivering an independent and objective design as proposed by the manifesto is unrealistic and biased, it has brought the profession away from social life, but this principle remains sustained in the architectural training in the country. As a result, working with and involving the communities in participatory processes are relatively new and challenging practices for local architects. Recently, they have been "forced" by massive housing needs to further explore and exercise these participation methods in the aftermath of three recent national-scale disasters (the Aceh tsunami in 2004, the Yogyakarta earthquake in 2006, and the Merapi eruption in

2010). Indonesian architects regained their social consciousness after the enormous destruction of the 2004 Aceh tsunami (local scholar, personal communication, February 2018).

In the following section, we shed light on three recovery projects implemented in the Sleman Regency after the 2010 Merapi eruption (see Fig. 7.2). Inhabitants from this regency are not only in danger of volcano eruptions, but also of earthquakes. The Sleman Government had made several previous attempts to relocate the inhabitants who were close to the summit of the volcano, yet always facing a fierce refusal from them (officer of a national public authority, personal communication, March 2018). The 2010 eruption provided the government with the leverage to evict them. The first recovery program was set up by Q-Tel following a top-down governance structure. Q-Tel constructed a relocation hamlet named Cancangan, housing around 60 families. The second program, called REKOMPAK (Community-Based

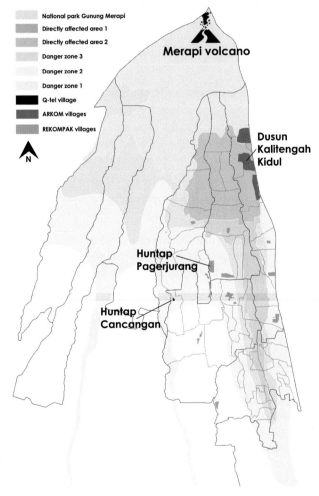

FIGURE 7.2 The location of the reconstructed villages of the three discussed programs in the Sleman District. *Credit: The authors*

Rehabilitation and Reconstruction Settlement Project), was designed by the Indonesian Government in a bottom-linked governance structure that favored relocation over rebuilding in situ. Under REKOMPAK, 12 hamlets were constructed in a safer location. This chapter focuses on the construction of the hamlet named Pagerjurang. The last program was set up by the community-architect group called ARKOM (Arsitek Komunitas, or Community Architects) in a bottom-up governance structure. It started with three communities refusing relocation and, as a result, being denied governmental support. Our study mainly sheds light on the reconstruction of the hamlet named Kalitengah Kidul. The following subsections discuss each case in more detail.

3.2 Q-Tel—Huntap Cancangan

Huntap Cancangan is a hamlet built by an international telecommunications company, named Q-Tel. Driven by a business vision to expand their market of telecommunications to Indonesia, Q-Tel entered the post-disaster site to conduct community work by offering complete houses on a relocation site to widows and families with young children through a lottery. Cancangan is the smallest hamlet out of all the new ones built after the disaster and is home to 58 households that originated from different villages before the eruption. Houses of 42 m^2 were constructed in eight months by a local contractor and all have the same design, with two bedrooms, a living room, a toilet, and a terrace with an outside kitchen. The hamlet benefits from basic facilities (such as a mosque, three little shops, and a playing area) but no other economic opportunities were provided for the beneficiaries. The local government in this instance was only involved in aiding to find a location for this new hamlet.

The foreign architect involved in this relocation project was hired and supervised by Q-Tel to design houses in a safe environment (see location in black in Fig. 7.2). Q-Tel urged the architect to act as a genius designer to get quick results, yet the architect attempted to familiarize himself with the local circumstances by gathering information about the beneficiaries' culture and by reaching out to a local Muslim boarding school. Nevertheless, the architect failed to set up community participation processes due to the lack of knowledge about who would be the final beneficiaries of the project at the time of construction. Therefore, the architect took the responsibility of designing the entire hamlet on his own.

3.2.1 Resilience building at the community level

Looking at the resilience-building process at the community scale, we uncovered an overall lack of cultural, social, economic, and political integration in the urban design through questionnaires with the inhabitants. The secluded *huntap* was not embedded into an existing urban fabric, which limits the access of the community to their families, markets, and other economic opportunities. Many inhabitants still perform jobs in their old hamlet (farming, volcano tourism, sand mining), which requires them to commute considerable distances every day. With the new inhabitants originating from 15 different villages, the hamlet has weak social cohesion and coping capacity. Moreover, the exclusion of the community from the layout design of the village generated multiple frustrations with respect to the offered facilities and infrastructure. The top-down governance structure established by Q-Tel and the architects' lack of knowledge on the local circumstances limited their potential to increase the overall resilience of the affected families at the community scale.

FIGURE 7.3 Houses with different finishing materials, which were applied by the inhabitants to differentiate their houses (head of hamlet, personal communication, March 2018) (situation in 2018). *Credit: The authors*

3.2.2 Resilience building at the housing level

The houses were constructed to be earthquake resistant, yet the structural resilience of the additions to the main house remains uncertain since the inhabitants did not receive any knowledge on how to build safe additions. Despite the architect's efforts to design the houses in a socially respectful way, he could not include the beneficiaries in the design processes as this was not possible following the rules set by Q-Tel. To address such failure, the inhabitants of the village made considerable changes to their housing layouts at a later stage of the recovery process (see Fig. 7.3). The urban fabric however does not allow many extensions, urging them to move elsewhere when the family grows.

3.3 The REKOMPAK program—Huntap Pagerjurang

Huntap Pagerjurang is one of the 12 hamlets constructed by the government (see the light gray areas in Fig. 7.2). It is the largest out of all the relocated villages, housing 301 households from five different communities. The site is located next to the Merapi golf course, where access roads were already built. Various facilities are situated across the site, such as mosques, playing areas, community buildings, and communal cattle sheds. Core houses of 36 m^2 were provided for each family (see Fig. 7.4), consisting of two bedrooms, a bathroom, an outdoor kitchen, and a living room. The community had the opportunity to choose the housing type from five different designs, while the internal layout of the house could be chosen by each family individually (architect, personal communication, February 2018).

Indonesian architects were recruited by the Indonesian Government and were included in the design process at various levels (national, district, and in situ). Additionally, the donor

FIGURE 7.4 Standard REKOMPAK house, owned by a widow with no financial means to make additions to it (2018). *Credit: The authors*

agency (World Bank) employed its architects to supervise the projects (national and district scales). Senior architects with previous experience in disaster sites (such as Aceh in 2004 and Yogyakarta in 2006) were hired to work at the national and district levels, and junior architects without previous experience were employed at the local level. The senior architects were responsible for drafting the codes and regulations and oversaw their implementation in the design and reconstruction process, gaining information from the junior architects. The junior architects worked closely with economists, sociologists, and engineers in a multidisciplinary team of facilitators to minimize possible negative repercussions of the relocation (for instance economic and social impoverishment). These junior architects also navigated between different stakeholders (various governmental agencies, World Bank representatives), active at different levels (local, district, national), to make the design process more inclusive (officer of a regional public authority, personal communication, March 2018). The junior architects took the role of active facilitators by setting up a two-way form of participation, making a possible knowledge transfer both from them to beneficiaries and the other way around (architect, personal communication, February 2018).

During the construction phase, the junior architects acted as building teachers and taught the local communities about safe construction techniques (architect, personal communication, February 2018). The senior architect responsible for overseeing the whole district devised the housing layouts (architect, personal communication, February 2018). A decision was made that every beneficiary would receive a similar house on an equally sized plot to minimize resentment among the community members. Based on equality and not on equity, such a decision did not pay attention to family sizes, the concept of extended families and vulnerable families. It further triggered debates and generated contradictions among the architects involved in this project in terms of their interpretation of equality, justice, and local habits.

3.3.1 Resilience building at the community level

Due to the extensive community-participation process aiming to embed social, cultural, economic, and political specificities into the urban design (officer of a local public authority, personal communication, March 2018), the implementation of this program was more successful in setting forward resilience-building processes at the community level as compared to the Q-Tel-designed one. The bottom-linked interactions between architects, other stakeholders, and donors were instated to make room for programming and policy improvements (officer of a national public authority, personal communication, March 2018), yet the implementation rules remained largely the same.

Moreover, the coproduced design was less successful in integrating new economic opportunities, according to a local scholar (personal communication, February 2018). The responses to our questionnaires also revealed that the inhabitants' livelihoods remained utterly dependent on activities connected to the Merapi Volcano (namely farming, sand mining, and volcano tourism). However, the overall economic status of the inhabitants did increase, because these activities flourished after the 2010 eruption (officer of a regional public authority, personal communication, March 2018).

3.3.2 Resilience building at the housing level

The senior architect who designed the house layouts was not present at the disaster site on a regular basis, but received information from the junior architects on the local specificities of the communities and implemented those into the design. Structural resilience was attained at the house level, whereby earthquake-resistant houses were constructed throughout the whole village (see Fig. 7.4). Every household was allowed to adapt the core structure of the new houses to foster the process of "home-making" (see Fig. 7.5), even though these adaptations were too limited in the case of extended families due to the small size of the plot. To ensure that the adaptations would follow the building regulations in force, the junior architects made an attempt to convey knowledge about safe construction to the community. However, what was observed in the buildings' extensions at the later stage of recovery was inhabitants' disobedience to the rules, which according to them was mostly due to financial reasons.

3.4 Community architects—Kalitengah Kidul

The inhabitants of three disaster-affected villages in the Sleman Regency (see the gray area in Fig. 7.2) refused to be relocated to a safer environment and, hence, to receive public financial support. Their refusal was grounded in the economic benefits of fertile lands and sand generated after volcano eruptions and also motivated by their belief of Merapi as God (inhabitant of Kalitengah Kidul, personal communication, March 2018). Determined to rebuild in situ, these inhabitants received support from the community-architect group called ARKOM, who resisted governmental orders due to their humanitarian standpoint that no person should be left out from the reconstruction process (architect, personal communication,

FIGURE 7.5 Expanded REKOMPAK house: The homeowners were expanding due to family growth (February–March 2018). *Credit: The authors*

February 2018). ARKOM already gained post-disaster experience in Aceh in 2004, when ARKOM's founders worked in the field for the Asian Coalition for Housing Rights and other NGOs while learning about community architecture. The Kalitengah Kidul hamlet, our research focus, was identified by ARKOM one month after the disaster. It is situated 4.5 km from the Merapi summit and is home to 110 families forming a tight community. The core structures were all built only three months after ARKOM's arrival and dimensioned according to the family's immediate needs (size, vulnerability of the family, etc.).

A multitude of architects' roles were combined in ARKOM's approach, the first of which was that of the compassionate friend, in their attempt to produce community-relevant design suggestions (architect, personal communication, February 2018). ARKOM also treated the community as the real architects and let them become the main decision-makers of their new living environment. ARKOM additionally played the role of the building teacher, providing the community with technical knowledge and suggestions on integrating disaster preparedness into the urban design. The pre-disaster conflicts and opposing views regarding relocation versus resettlement in situ came forward even stronger in the aftermath of the eruption. As a result, the village community was in urgent need of being supported by negotiators to change pre-disaster discourses and institutional practices. ARKOM took the role as intervening mediator, advocating for the community's best interest before various institutions, especially the local government, while encouraging the community to stand up for its rights (architect, personal communication, February 2018). The architects brought the strengths and capacities of the community to the fore of their negotiations.

3.4.1 Resilience building at the community level

The inhabitants of Kalitengah Kidul evidently held a high coping capacity already before the Merapi eruption. Merapi is a frequently erupting volcano, and the inhabitants of this hamlet have become highly sensible to understanding the natural signs indicating the pre-eruption situation (animals running away, wind, bamboo sounds, etc.). Yet, they persistently return to their hamlet after each eruption. ARKOM recognized these inhabitants' nature as an existing capacity and valorized it in several ways. They codesigned extensive emergency plans with the community to improve evacuation and decrease possible economic losses, namely adding cattle to the evacuation process, demarcating evacuation routes, and introducing the concept of evacuation bags with the most important documents. To strengthen the communities' claim to remain in their ancestral lands and to gather limited funds and technical assistance for infrastructure (such as sewage or water provision), ARKOM sought out strategic alliances on behalf of the community. They started a partnership with a local NGO to supply the community with spring water, and with WALHI (Indonesian Forum for Environment—Friends of the Earth), an organization striving for a just and democratic society who already had experience working with the Indonesian Government.

Eventually, the local government complied with various claims that ARKOM set forward on behalf of the community, namely provision of electricity, permission for teachers to work in the disaster-affected area, and cessation of relocation attempts (architect, personal communication, February 2018; inhabitant of Kalitengah Kidul, personal communication, March 2018). During the recovery process, ARKOM moreover contributed to expanding the public view of Indonesian architects who had been perceived narrowly and stereotypically as sterile and autonomous, by casting light on the openness of architects in familiarizing themselves with the local building culture and capitalizing on the communities' living knowledge.

3.4.2 Resilience building at the housing level

The houses in Kalitengah Kidul were rebuilt smaller due to insufficient funds, yet they allowed upgrading when individual budgets permitted it (see Figs. 7.6 and 7.7). It was a collective and rapid community effort and process. Within a period of three months, the core houses were designed and built by the families themselves and their neighbors, with limited technical aid from ARKOM (architect, personal communication, February 2018; inhabitant of Kalitengah Kidul, personal communication, March 2018). However, ARKOM already provided these core structures adapted to each family's needs by taking into account the family's size and vulnerability.

4. Community architecture and resilience building: Potential and limitations

In this section, we confront empirical evidences from Indonesia with our conceptual overview of the diverse roles of architecture in post-disaster rebuilding processes (see Table 7.1) with the aim to examine how architects' resilience potential is contingent to the governance structure of different reconstruction projects.

The governance structure of each recovery program largely determined the emergence of specific architect roles and, consequently, their resilience potential. The top-down governance

FIGURE 7.6   House in Kalitengah Kidul constructed in the aftermath of the 2010 Merapi eruption (situation in 2018). *Credit: The authors*

FIGURE 7.7   Upgraded house (situation in 2018). *Credit: The authors*

structure of the recovery project in Cancangan did not allow sufficient room for other roles to be taken up except for the genius architect, containing the resilience-building potential of the architect involved. Although Q-tel promoted the role of the genius designer in their architects' practice, it still allowed room for the architect to take the first steps toward a culturally and socially embedded design by consulting with the local institutions. The architect, however, did not make a real attempt to persuade Q-Tel for a more appropriate program implementation; instead, he complied with the imposed rules. Due to the relative rigidity of the recovery program, the architect did not become socially engaged and/or politicized, which limited the resilience potential of his practice.

The bottom-linked structure of the recovery project in Pagerjurang offered more space for participatory processes and multilevel interactions to take root, which widened the range of roles potentially activated by the involved architects. In this program, the facilitators worked under the regulations set up by the government without questioning the objectives or rationale of the program. As this program, although bottom-linked, was highly complex, with multiple governance levels and a rigid structure, little room was left for socially innovative practices to take root by the engaged architects. The claims inhabitants conveyed through the junior architects and other members of the team of facilitators were seldom effectively communicated to higher levels of the government and thus not implemented.

Lastly, the bottom-up structure of the recovery project in Kalitengah Kidul enabled the emergence of a wider range of architects' roles, which were also more responsive to the specific needs and capacities of the community leading the reconstruction project. The bottom-up governance modality allowed the community to take control over the design of their hamlet, while still receiving technical and advocacy aid from ARKOM to rebuild a more socially relevant post-disaster urban settlement.

5. Conclusions

This research revealed that a multitude of different roles—single or combined—are taken up by architects who participate in reconstruction initiatives. The extent to which the chosen roles contribute to building resilience and reducing vulnerability in the contexts of heterogeneous disaster-affected communities depends largely on the following: the architect's understanding, respect, and incorporation of the cultural, political, and economic specificities of each intervention area; the manifestation of a multiplicity of roles architects should manifest; and the governance structure of each reconstruction project and its openness in bolstering multilevel interactions.

An important finding in our empirical research was that the role of the radical reformer as anticipated in our conceptual overview was not evident. The absence of architects as activists challenging hegemonic power and institutional rigidities, and encouraging spaces of negotiation and contestation restricted the accomplishment of the profession's full potential in improving governance structures and building resilient communities in a comprehensive manner.

How can architects overcome governance rigidities to bolster and optimize the potential of their profession to building resilience in disaster-recovery contexts? We argue that the institutional and programming rigidities can only be addressed when architects become radically politicized and connected with larger sociopolitical movements. Future research should shed more light on the political roles architects (can) play to overcome the current institutional constraints in post-disaster contexts and thus guide an overall recovery project that benefits all the affected communities.

Acknowledgements

The first author is grateful to VLIR-OUS for the travel grant to conduct fieldwork in Indonesia.

References

Aquilino, M. (2011). Beyond shelter: Architecture and human dignity. In M. Aquilino (Ed.), *Beyond shelter: Architecture for crisis* (pp. 6—10). London: Thames and Hudson.

Architecture for Humanity. (2012). In *Design like you give a damn [2.* New York, NY: Abrams.

Boano, C., & García, M. (2011). Lost in translation? The challenges of an equitable post-disaster reconstruction process: Lessons from Chile. *Environmental Hazards, 10*(3—4), 293—309. https://doi.org/10.1080/17477891.2011.594493.

Boano, C., & Hunter, W. (2012). Architecture at risk (?): The ambivalent nature of post-disaster practice. *Architectonica, 1*(1), 1—13. https://doi.org/10.5618/arch.2012.v1.n1.1.

Boano, C., & Talocci, G. (2017). Inoperative design: 'Not doing' and the experience of the community architects network. *City, 21*(6), 860—871. https://doi.org/10.1080/13604813.2017.1412649.

Charlesworth, E. (2007). *Architects without frontiers: War, reconstruction and design responsibility.* London: Architectural Press. https://doi.org/10.4324/9780080465449.

Charlesworth, E. (2014). *Humanitarian architecture: 15 stories of architects working after disaster.* New York, NY: Routledge. https://doi.org/10.1017/S135913551400061X.

Coulombel, P. (2011). Open letter to architects, engineers and urbanists. In M. Aquilino (Ed.), *Beyond shelter: Architecture for crisis* (pp. 286—295). London: Thames and Hudson.

Davoudi, S. (2012). Resilience: A bridging concept or a dead end? *Planning Theory & Practice, 13*(2), 299—333. https://doi.org/10.1080/14649357.2012.677124.

Eizaguirre, S., Pradel, M., Terronesm, A., Martinez-Celorrio, X., & Garcia, M. (2012). Multilevel governance and social cohesion: Bringing back conflict in citizenship practices. *Urban Studies, 49*(9), 1999—2016. https://doi.org/10.1177/0042098012444890.

Habraken, N. J., & Teicher, J. (1972). *Supports: An alternative to mass housing.* London: Architectural Press.

Jigyasu, R. (2008). Structural adaptation in South Asia: Learning lessons from tradition. In L. Bosher (Ed.), *Hazards and the built environment: Attaining built-in resilience* (pp. 74—95). London: Routledge.

Kusno, A. (2012). *Zaman baru generasi modernis: Sebuah catatan arsitektur [The modernist generation of the new age: An architectural note].* Yogyakarta, Indonesia: Ombak (in Indonesian).

Lorenz, D. F. (2013). The diversity of resilience: Contributions from a social science perspective. *Natural Hazards, 67*(7), 7—24.

Luansang, C., Boonmahathanakorn, S., & Domingo-Price, M. L. (2012). The role of community architects in upgrading: Reflecting on the experience in Asia. *Environment and Urbanization, 24*(2), 497—512. https://doi.org/10.1177/0956247812456125.

Maly, E., Iuchi, K., & Nareswari, A. (2015). Community-based housing reconstruction and relocation: REKOMPAK program after the 2010 eruption of Mt. Merapi, Indonesia. *Institute of Social Safety Science Journal, 27*, 205—214.

Manyena, B., O'Brien, G., O'Keefe, P., & Rose, J. (2011). Disaster resilience: A bounce back or bounce forward ability? *Local Environment: The International Journal of Justice and Sustainability, 16*(5), 417—424. https://doi.org/10.1080/13549839.2011.583049.

Oliver-Smith, A. (1991). Successes and failures in post-disaster resettlement. *Disasters, 15*(1), 12—23. https://doi.org/10.1111/j.1467-7717.1991.tb00423.x.

Paidakaki, A., & Moulaert, F. (2017). Does the post-disaster resilient city really exist? A critical analysis of the heterogeneous transformative capacities of housing reconstruction 'resilience cells'. *International Journal of Disaster Resilience in the Built Environment, 8*(3), 275—291. https://doi.org/10.1108/IJDRBE-10-2015-0052.

Paidakaki, A., & Moulaert, F. (2018). Disaster resilience into which direction(s)? Competing discursive and material practices in post-Katrina New Orleans. *Housing, Theory and Society, 35*(4), 432—454. https://doi.org/10.1080/14036096.2017.1308434.

Pelling, M., & Dill, K. (2010). Disaster politics: Tipping points for change in the adaptation of sociopolitical regimes. *Progress in Human Geography, 34*(1), 21—37. https://doi.org/10.1177/0309132509105004.

Samekto, C. B. D., & Nuh, M. (2017). Evaluation of community-based settlement reconstruction program (case study in post-disaster recovery of 2010 Merapi Volcano eruption in Cangkringan district, Sleman regency, Yogyakarta special region). *Journal of Public Administration Studies, 1*(3), 64–70.

Tauber, G. (2015). Architects and post-disaster housing: A comparative study in South India. *International Journal of Disaster Resilience in the Built Environment, 6*(2), 206–224. https://doi.org/10.1108/IJDRBE-07-2013-0025.

Turner, J. F. (1972). The reeducation of a professional. In J. F. Turner, & R. Fichter (Eds.), *Freedom to build: Dweller control of the housing process* (pp. 122–147). New York, NY: Collier Macmillan.

Turner, J. F. (1980). What to do about housing—its part in another development. *Habitat International, 5*(1–2), 203–211. https://doi.org/10.1016/0197-3975(80)90074-0.

Wates, N., & Knevitt, C. (1987). *Community architecture: How people are creating their own environment.* London: Penguin.

Zetter, R., & Boano, C. (2009). Space and place after natural disasters and forced displacement. In G. Lizarralde, C. Johnson, & C. Davidson (Eds.), *Rebuilding after disasters: From emergency to sustainability* (pp. 206–230). London: Routledge.

8

Probing for resilience: Exploring design with empathy in Zanzibar, Tanzania

Helena Sandman, Miia Suomela

Aalto University, Helsinki, Finland

1. Introduction

Sustainable Development Goal 11—making "cities and human settlements inclusive, safe, resilient and sustainable" (United Nations, 2018a)—constitutes a major challenge in developing countries. In these contexts, rapid urbanization, coupled with the absence or ineffectiveness of local housing policies, has resulted in an increasing number of people living in informal settlements that are prone to disasters and add to urban sprawl. This development

Enhancing Disaster Preparedness
https://doi.org/10.1016/B978-0-12-819078-4.00008-3

is neither inclusive, safe, nor resilient. The vulnerability of both the environment and the inhabitants increases the need for focusing on resilience.

In this regard, resilience is widely defined as the capacity of a system, community, or society to resist and absorb disturbance, to adapt to change, and to transform while maintaining its core characteristics and continuing to develop (Stockholm Resilience Center, 2015; UNISDR, 2017; Walker & Salt, 2006). Walker and Salt (2006) consider resilience as a key to achieving sustainability in *social-ecological systems*—complex integrated systems in which humans and nature are bonded into a whole. They argue that the more resilient a system is, the better it can provide services essential to life, which can be considered a characteristic of sustainability. Resilience emphasizes the importance of viewing the system as a unity instead of breaking it into smaller parts that are considered independently. Walker and Salt (2006) suggest that partial solutions in isolated components of the system may eventually result in more serious problems, which can be reflected in other spatial or temporal scales.

Social scientists have criticized the concept of social-ecological systems for "undertheorizing" the involved social entities; researchers in the field are debating whether society is too complex an entity to be conceptualized as a component of social-ecological systems (Stojanovic et al., 2016). We argue, however, that there are benefits considering communities in developing countries as unified social-ecological systems, stressing the connection between society and the ecological realm. We think that it is important to emphasize social aspects when striving for sustainable societies to enable, for example, human health and well-being, affordability, and cultural preservation in a community. Nevertheless, we claim that it is possible to achieve these while also protecting the environment and ensuring the future provision of ecosystem services that the social realm depends on. When social and ecological aspects are considered equally important components of a social-ecological system, they reinforce one another and support the resilience of the system.

A resilient social-ecological system, through the contribution of its active members, has the capacity of turning disturbances into opportunities for innovation and development (Folke, 2006). To this end, as noted by the architect Charles Correa (1994), people's engaged participation is an essential aspect of our habitat. In this regard, one of the core principles of resilience building, according to authors associated with the Stockholm Resilience Centre, is "broadening participation" to address both social and ecological aspects of improving resilience (Simonsen et al., 2015). In previous research, Sandman, Levänen, and Savela (2018) argued that inhabitants' engagement in the architectural design process plays an important role when striving for sustainable societies. Likewise, we assume that resilience can be improved by the engagement of people by architects in the formal processes of spatial change. When community members are effectively engaged in the development of their habitat, they are empowered and inspired to embrace endeavors that promote the community's resilience, which further improves the system's sustainability. In this chapter, we refer to resilience from the perspective of urban development and a community's adaptation to change, and not particularly to resilience related to disasters.

In a design process, when engagement is guided thoughtfully with sensitivity, it enhances the relationships between stakeholders and builds trust and a shared understanding (Akama & Yee, 2016; Mattelmäki, Vaajakallio, & Koskinen, 2014). Yet, architects would need to build their professional capacity to meet the challenges of engaging inhabitants to improve resilience in rapidly evolving societies. In developing countries, where there might be insufficient professional resources and where stakeholder engagement can be challenging, future inhabitants might not be empowered, might not have the time and energy to invest in the project, or might not be accustomed to taking part in a design process. In these contexts, entanglements, obstacles, or gaps between stakeholders often appear in architectural design projects (Hussain, Sanders, & Steinert, 2012). Thus, approaches that bring attention to bridging these gaps are needed. We hypothesize that a way to surmount these constraints is to enhance empathy and understanding between people—architects and community members alike. If the design process is conducted with empathy, we argue that it supports resilience building.

Empathy can be defined in many ways. Most theorists agree that empathy is the ability to "ascribe mental states to others," taking the perspective of another, or the process of being affected by another person's emotions (Maibom, 2017, p. 1). Here, we refer to empathy broadly as experiencing and appraising the world from another's point of view, not only with an emphasis on emotions. Experiencing the world naturally involves emotional states, but it also involves practical, habitual, and cultural components. As architects seldom design for themselves, this ability ought to be a core competence in the profession. The importance of empathy is even greater in developing-country settings, where the clients may be in vulnerable situations due to prevailing sociocultural structures (Hussain et al., 2012). As such, we argue that without empathic engagement it can, for instance, be difficult to identify social factors and local architectural features to be honored in the design process. Both are important aspects if the intention is to maintain and foster resilience and sustainability.

Through our interest in empathy, we paid attention to the empathic-design discourse. Empathic design is an approach that encourages empathic understanding between designers and users (Koskinen & Battarbee, 2003). It suggests that the designer should have an open-minded attitude, observational skills, and curiosity (Leonard & Rayport, 1997). We studied this approach originating in product design to identify possible applications in the field of architecture, given that product-design processes are often faster and more flexible than architectural ones. Indeed, the latter can be long, substantial, and heterogeneous, due to an extensive number of stakeholders and legal codes governing these practices (Mazé, 2007).

Empathic engagement in architecture in developing countries is the central topic of the first author's doctoral research at Aalto University, Finland. This chapter presents one of the aspects covered under this broader topic. Here, we focus on *design probing*, a method utilized in empathic design and participatory design, which can encourage multiple ways of empathic engagement. It has been promoted as a valuable tool when designing in socially critical contexts with and for vulnerable users (Debrah et al., 2017). Through two design studies developed in Zanzibar, Tanzania—one of the world's least developed countries (United Nations, 2018b)—, we illustrate how design probing can support the work of architects in developing countries. Focusing on these contexts, we examine the advantages of design probing to broaden participation within architectural design. As such, this chapter presents two ways of employing design probing and discusses its practical use. We elaborate on the experiences and benefits of the method and demonstrate how the probing exercises can inspire and inform the design, support personalized local solutions, and enforce empathic engagement.

2. Broadening participation through empathic and creative methods

Broadening participation improves legitimacy, increases knowledge, and helps detect disruptions; in particular, early engagement helps in defining priorities and needs (Simonsen et al., 2015). Moreover, as regards architectural design processes, broadening participation can assign the inhabitants an active role to reach results suitable to their actual needs. To support their engagement with inhabitants, architects need to find efficient, yet empathic, methods suitable for challenging situations in fast-growing cities in the developing world, when there are time constraints and other limitations. In the architecture literature, not much is written on alternative methods for inhabitants' empathic engagement, whereas in the context of design research, the concept of empathy and empathic-design methods are well known.

In our previous research, we have gained insight into the potential of an empathic approach to architectural design (Sandman, Maguire, & Levänen, 2020). We identified four complementing registers of empathic understanding and engagement within the design process. Firstly, the architects' imagination plays a strong role. Architects imagine themselves as inhabitants or users of a space (Pallasmaa, 2015). Secondly, architects observe the users' lives, their habits, and activities (Steen, 2011). Thirdly, architects involve users in the design process in a sensitive manner and ask them to share their views, thoughts, and dreams (Koskinen & Battarbee, 2003; Mattelmäki et al., 2014). Fourthly, architects meet users on an intimate level and reflect together on similarities and differences in their experiences (Akama & Yee, 2016). In this fourth option of empathic engagement, all registers merge and deepen.

In the product-design discipline, multiple methods and approaches of empathic design have been developed during the last three decades (Mattelmäki et al., 2014; Sanders & Stappers, 2008; Steen, 2011). Whereas the definition of empathic design is broad and includes a variety of methods, what is common to all of them is the foundation of curiosity and willingness to step into other people's shoes (Koskinen & Battarbee, 2003). Furthermore, empathic design is defined as an approach that focuses on what ought to be (Steen, 2011). This falls back on the philosopher David Hume's well-known is—ought problem: What ought to be cannot be based on what is (Hume, 1739/1896). Therefore, methods that exclusively study what is are not enough in a design context, in which the aim is to create something new.

Design probing is an empathic-design method that has proved its value as part of a larger participatory agenda. It is a tool for understanding human phenomena and unveiling design opportunities (Mattelmäki, 2006). Mattelmäki (2006) described probes in her doctoral thesis through three features: the focus of the assignments on the user's perspective in a broad sense—from the cultural environment to feelings and needs—, the participants' self-documentation, and the exploratory character of the exercise, seeking to identify new opportunities. Consequently, the emphasis of probes is to inspire what ought to be, in contrast to capturing what is (Boehner, Gaver, & Boucher, 2012); between the *is* and the *ought to be*, there is space for creativity. In this respect, design probes are meant to support both users and designers in expanding their creativity. Undeniably, creativity is the main driver in the different phases of the probing process. Firstly, designers create the probes as inspiring as possible to be distributed to participants who, secondly, creatively accomplish the tasks, and thirdly, designers utilize the material received from participants as creative inspiration for the design task (Gaver, Dunne, & Pacenti, 1999). The motivator for creativity, in this case, is empathy, as all the phases of the probing exercises concern the experience of the users.

The registers of empathic engagement and understanding, previously presented, can also be detected throughout the entire process of design probing. To design the probes, designers imagine themselves in the place of users, based on their own experiences. In this stage, the capacity to involve personal experiences to deepen the imagination is valuable (Pallasmaa, 2015; Smeenk, Tomico, & Van Turnhout, 2016). In the second stage, designers create inspiring tasks for users to let them share important aspects of their lives. Here, designers engage with users with sensitivity to be able to reach their emotions and aspirations (Gaver, Boucher, Pennington, & Walker, 2004; Mattelmäki, 2006). When applying design probing, there is always uncertainty; it is not possible to know what responses will be received, as the intention is not to guide the participants in any sense. This aspect that honors uncertainty also requires a sensibility from designers who utilize probing (Boehner et al., 2012). Moreover, when the users receive the probes and are confronted with their design features, they can obtain an intimate insight into the creativity of the designer.

In the third stage, designers seek to understand the responses empathetically, not merely intellectually (Gaver et al., 2004). Here, the aim is to bridge the gap between the stakeholders identifying similarities and recognizing differences in their understandings and experiences. In this stage, a relationship on an intimate level might be established between designers and users. This is possible even if a distance between them is inevitable. The probes tend to "create relationships [between designers and participants] that are a little like designing for friends: We know them well" (Gaver et al., 2004, p. 6). Therefore, probing can be perceived as part of an ongoing empathic dialogue that nurtures understanding between designers and the people and places they are designing for (Boehner et al., 2012).

Furthermore, the designer, when receiving back the probes, acquires an opportunity to be part of certain aspects of the users' lives that would have stayed obscured otherwise, due to the distance between them. As pioneers in design methods and probing, the psychologist and design researcher William Gaver et al. (2004) explained that for the participants, the activity can make the familiar seem interesting when viewed through different lenses. On the other hand, for the designer, it can illustrate something peculiar and through personal insight make it familiar (Gaver et al., 2004). The participants, upon receiving the probes, do not know the exact intention behind the exercises because of the distance between them and the designer. Thus, they can personally interpret the exercises and respond with creative freedom. Likewise, for the designer, this detached, still close, view into somebody's life can be a fruitful standpoint for innovative design ideas; such an *intimate distance* leaves the freedom required for creativity (Gaver et al., 2004).

Regardless of the seemingly open approach, there is within the design discourse a discussion about the purpose of probing. Gaver et al. (2004) criticized the application of probing for obtaining information instead of getting inspiration. They argued that applying probes to get objective answers in research frameworks endangers the original intentions of the method, which values uncertainty, play, and exploration. Furthermore, they argued that most research techniques tend to disguise subjectivity through controlled procedures, whose results can be considered impersonal whereas probes take the opposite approach. In their own probing processes as designers, Gaver et al. (2004) refrained from believing that they could scrutinize the heads of the users and instead made use of their subjective interpretations (Gaver et al., 2004). Encouraging this subjective engagement and empathic interpretations, Gaver et al. (2004, p. 56) still conceded that probes can be used for collecting research materials; however, they anticipated that probes' original motivation, to retain a "pervasive sense of uncertainty," should be respected. We thus explored in our research variations of this method to contribute to empathic engagement in architectural design processes in developing-country settings.

3. Experimenting with design probing

We were introduced to Zanzibar through Dr. Muhammad Juma, the director of the Department of Urban and Rural Planning (DoURP) in 2014 while the first author was teaching the Aalto University's master course Cities in Transition in Dar es Salaam, Tanzania. When we learned about the DoURP director's aspirations for sustainability and the concerns regarding insufficient numbers of professionals, we decided this place would be the focus of both Helena's doctoral studies and the subsequent edition of the course. In 2018, the course, then renamed Interplay of Cultures, started to collaborate with the DoURP to engage with urban challenges in Zanzibar Town. These issues include the need to accommodate more inhabitants in the central parts of the city and to plan for sustainable new areas as urbanization is accelerating. The department is concerned with the risk of losing intangible cultural heritage if uncontrolled development forces present inhabitants to move to the outskirts of the town. Additionally, unrestrained urban sprawl has been encroaching on valuable agricultural land, which is a threat to the densely populated island (Juma, 2014).

During our collaboration with the DoURP in Zanzibar, we utilized design probing as a method to engage with communities at the beginning of the design process. In this chapter, we present two examples in which probing was used in different ways. The first example is an urban-design exercise for the Chuini neighborhood on the northern outskirts of Zanzibar Town, carried out by Miia Suomela for her master's thesis in architecture, as a continuation of the Interplay of Cultures course that she attended in 2018. The master plan for Zanzibar proposed that Chuini be developed into one of six subcenters to ease pressure from the city center. In this case, the probing was conducted as an inspirational exercise motivated primarily by ecological sustainability due to the environmental vulnerability of the area.

The second example is the densification and affordable-housing design in the Ng'ambo neighborhood of central Zanzibar Town undertaken as part of Helena Sandman's aforementioned doctoral thesis. In this case, the design probes were motivated by social sustainability, and the main aim was to gather information about the inhabitants' perception of their homes. In both exercises, design probing was only one of the participatory methods applied in the design and was executed at the beginning of the process to initiate contact with the community on a personal level. By *personal*, we refer to both the promoted face-to-face meetings between individuals and ourselves and the intention that the inhabitants share their individual views without being influenced by their families or neighbors, as there might be unknown hierarchical levels within these groupings.

As we used design probing for empathic engagement and for bringing stakeholders closer to each other, we also want to clarify that in our case the distance was cultural, linguistic, and geographic, as we came from a different part of the world. The qualitative approach of our research was interpretive and subjective and took advantage of embodied and situated knowledge while recognizing existing limitations. For instance, our knowledge of the Swahili language was limited, and therefore some of the discussions were conducted with the aid of a local research assistant and a young architect of the DoURP.

3.1 Probing for ecologically resilient urban design in Chuini

In the first design-probing exercise, we focused on ecological sustainability and sought to get inspiration for the design. The selected site, the Chuini neighborhood, is characterized by rapidly expanding informal settlements, agricultural activities, proximity to the ocean, and lush greenery. The master plan for Zanzibar Town assigned Chuini to be developed mainly into a residential area and proposed it to host ninefold its current population of 10,000 people in 2035. Given that all existing agricultural land, mostly wetland, is proposed to be sacrificed for development, the planning exercise in Chuini focused primarily on ecological resilience.

The DoURP is studying alternative patterns of densification in Chuini, aiming to develop the region in such a way as to preserve the greenery and to maintain and enhance its ability to retain stormwater and mitigate floods. Preserving the agricultural activities is crucial for the ecological resilience of Chuini but, as a source of food and income, they are also a vital component of the community's socioeconomic resilience. Given the fragility of this social-ecological system, the challenge is to respond to the urgent densification needs in a socially sustainable way while protecting the environment and ensuring the future provision of ecosystem services.

The design probing aimed to gain an understanding of the experiences and thoughts of the inhabitants of Chuini regarding their environment. We assembled a simple probe kit that included a card with an introduction of the master's thesis, instructions on how to proceed with the kit, a pen, and three packs of cards, each of them including five postcard-sized pieces. On the front of each card, there was a picture and on its back a question, in English and Swahili, and some blank space for writing or drawing an answer. The first set of cards asked: "When you think of your neighborhood what comes to your mind first when you imagine the color green/blue/red/yellow/white?" The second set asked: "When you think of Chuini what comes to your mind first when you consider the word city/house/home/people/water?" And the third set asked: "When you think of your everyday life what comes to your mind first when you look at the picture on the other side of this card?" The corresponding photos are presented in Fig. 8.1. In addition, the participants were asked to tell their age, gender, and occupation. These exercises aimed at identifying the associations that a set of colors, words, and pictures would awaken in the participants' minds, which could then constitute sources of inspiration in the design process.

A local *sheha* (the head of a *shehia*, the smallest administrative unit in Zanzibar) chose 15 households to participate in this probing exercise. After deciding that, to avoid excessive peer influence, it was better to deliver the probe kits individually to each household than to deliver them in a group meeting, we walked from house to house together with the *sheha*'s assistant and our research assistant who translated the discussions. We offered a kit to each participant, gave an overview of its contents, and explained why we wanted them to participate in such an exercise. Despite our intentions, we did not get to meet all the participants personally, because four kits were left with the *sheha* to be delivered to inhabitants living a little further away.

We allowed 5 days for the participants to fill in the cards and return them to the *sheha*, from whom we picked them up after the deadline. Altogether 10 kits were returned from three women and five men aged 40−68, and two participants who did not share their details. The involved *sheha* probably encouraged the participants to respond to the probes but

FIGURE 8.1   Pictures used to invoke associations with everyday life. From left to right: tree crown, a muddy puddle, plastic bottles in an open drainage, baked bricks, and colorful fabrics. *Credit: Miia Suomela (2019).*

also might have put pressure on them to answer in a certain way. It is difficult to estimate the impact of the *sheha*'s involvement, but we worked on the assumption that the participants responded to the questions individually and uninfluenced as we asked them to.

The answers provided valuable insights for us into the thoughts and feelings of the inhabitants of Chuini, regardless of the shortcomings. All except one respondent answered in Swahili. We carefully studied the responses but did not summarize or count them to emphasize their role as inspiration instead of information. A couple of the original responses are portrayed in Fig. 8.2. In those cards, yellow is depicted as a color "that shows a good beginning of the day in the morning" and red sparks the thought: "We condemn with all force the lack of peace inside our Chuini." The word "people" makes one participant think about poverty and unemployment and to conclude: "We need to be empowered." The word "house" inspires a very universal thought: "Every human being wants a place to live." The picture of colorful fabrics makes a participant express their hope: "I wish that there was a factory that would make fabrics for making clothes." The picture of plastic bottles in a drain spurs a call for action: "I think that we should take care of our environment."

In the other responses, the words made the participants express their concern for the unplanned urban sprawl in Chuini and how it has destroyed the natural environment and invaded cultivated land. Regarding colors, green and blue are associated with nature, which the participants value as an essential part of their environment and would like it to be preserved. The photos sparked hope of modern homes and services to be available. Besides, participants addressed the preservation of livelihoods and the creation of employment opportunities. The probe responses inspired the urban design of future Chuini mandating the prioritization of protecting green areas while considering socioeconomic aspects. Thus, our design proposed the preservation and enhancement of most of the green areas, the structuring of and better connections to the residential areas, and versatile social services and economic activities, as suggested by the participants.

FIGURE 8.2 Some of the original probe responses in Swahili. *Credit: Miia Suomela (2019).*

3.2 Probing for socially resilient housing design in Ng'ambo

In the second design-probing exercise, we focused on social sustainability and sought mainly to gather information about the community, even if the results also inspired the design. Together with the DoURP, we chose for our affordable-housing design an inner-city site with 13 houses and approximately 100 inhabitants in Ng'ambo, a predominantly low-rise neighborhood. Many of the inhabitants had lived in the area for generations and some of them had lost their houses due to the construction since the 1970s of the Michenzani apartment blocks: 12 seven-floor high 300-m long buildings in the core of their neighborhood (Folkers, 2014). This major change in Ng'ambo was still fresh in the inhabitants' memories. Thus, they were aware of the risk of eviction they might face, which made the task of establishing trust between the involved stakeholders a challenge. For the design-probing exercise, we chose to engage only 5 of the 13 families because we wanted to make the probe packages rather extensive, with a focus on quality instead of quantity. Moreover, probing was only one of the participatory exercises we wanted to conduct, and we did not want to exhaust all 13 families at the beginning of the design process, considering their possible time constraints.

The probing exercise intended to make the inhabitants reflect on their relationships with their homes and to encourage them to observe their surroundings. The probing package contained artifacts and exercises designed to enable participants to illustrate what daily life is like in Ng'ambo. We strived to make the probing package personal and yet familiar, using locally available material (Fig. 8.3). As we were applying design probing in an architectural project for the first time, we crafted the probe package in a rather neutral way in order not to risk disturbing the results by the material choices, opting for a handcrafted style. The exercises were thoroughly explained in the package and designed to be concrete.

With stickers, a disposable camera, and pens included in the probing package, the participants had to accomplish the following exercises:

FIGURE 8.3   The design-probing package. *Credit: Helena Sandman (2016).*

— mark with different colored stickers items or parts of the house that they either favored or disfavored;
— take pictures of the placed stickers;
— take pictures of the places visited and the people met during the exercise timeframe;
— draw a map of places visited during the exercise timeframe;
— draw a plan of their present and their dream houses; and
— reply to a few questions regarding their life in Ng'ambo.

Through our discussions before and after accomplishing the exercises, the participants also received information about the plans for the area.

Introducing the probing exercise required personal contact to create trust and an appropriate framing of the situation (Fig. 8.4). In each household, according to local customs, we asked the oldest person to choose who would assume the task of the probes, a practice that led to a natural inclusion of different generations, thereby yielding a variety of views. The elder generation was also involved in the meetings at the beginning and at the end of the exercise, whereas the chosen probers (three women and two men) were all young adults, except for one, who was the head of the house and a single mother. During the distribution of the packages, we introduced the project and went through the exercises in detail, and allowed the participants two weeks to complete the probes.

Subsequently, in each household, we had a thorough discussion about the exercises and the replies. According to the feedback received, the participants enjoyed doing the exercises and found it interesting to reflect on their relationship to their home and their neighborhood. In the outcomes, on an informative level, we noticed a desire for new modern spaces and furniture and a dislike of the worn-out parts of the buildings and broken furniture. Similarly, the inhabitants also criticized items that consumed much electricity, due to high costs and frequent power cuts. Additionally, they revealed a wish for better sanitation and functioning infrastructure. Two of the participants wished for more privacy, particularly concerning the toilet and bathroom spaces. The responses also indicated a lack of proper cross-ventilation in

FIGURE 8.4 Our research assistant Saada, on the right, introduces the probing package to Mwanakombo. *Credit: Helena Sandman (2016).*

the houses (Fig. 8.5). The photos taken by the participants showed how they spent their time and what parts of the home drew their attention. The exercises pointed out some spots in the neighborhood that were important (Fig. 8.6) and made it clear that outdoor life, green spaces, and vegetation were appreciated.

The floor plans of the houses drawn by the inhabitants were informative as it was interesting to compare these with the actual plans of their houses, which were previously measured. For instance, when space was perceived as good or important, it was often drawn bigger than its actual size, whereas when disliked it was drawn smaller. In two cases, the room of the participant was bigger than its actual size, whereas in four plans the bathroom was drawn much bigger than it was in reality. In two drawings, the living room and the

FIGURE 8.5 Ali disliked the fact that one of the windows in his room was closed and prevented cross-ventilation due to an extension of his house. *Credit: Helena Sandman (2016).*

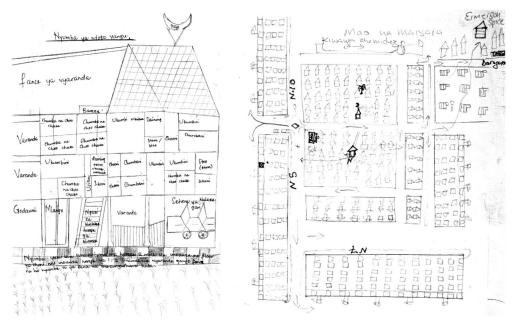

FIGURE 8.6   Two completed exercises showing a dream house and important spots in the neighborhood. *Credit: Helena Sandman (2016).*

veranda (and in a case also the backyard), the common places of the home, were also drawn bigger than in reality. These plans essentially illustrated how the participants perceived their homes, which might be more important than factual measurements for a design project. It was also interesting to compare the drawings of their existing homes with those of their dream homes. Some of the dream homes were close to copies of their existing ones. However, three dream homes demonstrated that living on multiple floors would be preferable to living in one-story houses. This fact is encouraging, for higher buildings are necessary when densifying an existing urban structure.

Through the probes, we also learned that people generally agreed that life in Ng'ambo was peaceful and nice, and its inhabitants took advantage of living in the city center. Three participants praised the social connections and the quality of knowing your neighbors, whereas two preferred more privacy. Additionally, we identified significant information regarding the use of space, social factors, and architectural features of importance. Furthermore, it proved an eye-opening experience to discuss the exercises and reflect on the concept of home with the participants and their families. Thus, the actual housing design was strongly influenced by the probing exercise, which generated solutions that neither we nor the participants would have been able to create in isolation. The rich materials inspired us to design a housing solution that would leave an opportunity for the inhabitants to develop their own homes incrementally in a personal way (Fig. 8.7). We designed a three-floor building around a courtyard, with 24 apartments. Each apartment could consist of a varying number of rooms, depending on the needs of each family (Fig. 8.8).

FIGURE 8.7   The flexible floor plan and façade of the resulting affordable-housing proposal for Ng'ambo. *Credit: Helena Sandman, Mariana Rantanen, & Ivan Segato (2017).*

Through the responses, we understood that the apartments required proper cross-ventilation and a space akin to an open courtyard, where laundry and kitchen activities could take place. Parts of the apartments could be left open, as large outdoor areas. Moreover, the inhabitants could choose themselves how much space would be utilized for indoor rooms and the outdoor terrace. In this respect, we intended to leave the same amount of flexibility for the inhabitants within the walls of the apartment, as they have now in their private houses, where the plot border constitutes the limits. We also understood the value of outdoor social meeting points (*barazas*), which take the form of either a veranda next to the house's main entrance or a group of benches around a tree in a public spot in the neighborhood where men often gather in the evenings. This possibility to sit outdoors for a chat with the neighbors was also a feature added to the design due to the probing exercise. To accommodate this, we suggested *barazas* along the streets, in the courtyard, and the open staircases.

4. Discussion: Engaging through inspiration and information

The probing exercises in Ng'ambo and Chuini demonstrated contentment with the present situation on some levels and a positive attitude toward the urban development to come, yet we also identified suggestive patterns of concerns and wishes. Although in the two cases the

FIGURE 8.8   A view of the proposed courtyard. *Credit: Helena Sandman, Mariana Rantanen, & Ivan Segato (2017).*

probing exercises aimed at either getting inspiration or gathering information, our experiences indicated that probes in architectural design are helpful in both dimensions. Through the conducted experiments, we consider that a combination of information and inspiration is indeed the most fruitful.

After these experiences, we would be much more flexible regarding the design of the probe kit. For instance, we wanted to use a disposable camera in the exercises in Ng'ambo, as this is a typical probing tool and we did not know beforehand how common the use of smartphones was in Zanzibar. However, we learned that disposable cameras are not suitable for a probing exercise. Firstly, they are not sustainable, secondly, the quality of the pictures is poor, and thirdly the item was strange and unfamiliar to the participants. We found out that smartphones are very common among the inhabitants of Ng'ambo and could be easily used in a probing exercise. Nevertheless, our design skills should always prevail when crafting the probes, regardless of the resort to digital possibilities. When the probes are produced with intention and care, the exercise package becomes something nice to receive and, consequently, the replies are likely given equal care, as in the probing results we received.

Furthermore, the visual impression of the probe package might have stirred the participants' imagination by encouraging them to look at ordinary things from a new perspective, as revealed for instance by the poetic comments on the cards in the Chuini case or the creative drawings of dream houses in the Ng'ambo case. In the latter, when we received and studied the probing results, it felt as though we had been visitors in the households for a much longer period than the brief introductory meetings that we had on the porches. It also made us feel like guests, bringing to our attention aspects of the homes that would have stayed obscured through observations alone. We came to appreciate these homes and their inhabitants. The design probing opened doors to the lives of the participants, which otherwise would have been challenging to access given the available resources.

The responses further implied that the community members are open to various spatial possibilities and willing to participate in the design process. The personal reflections revealed in the probes made us deeply grateful to the participants. We were touched by the openness and trust the participants showed us. In the two cases, we experienced the flexibility and versatility of design probing, which yielded deep insights into the inhabitants' world without demanding excessive efforts on either part. Through these experiences, we could easily agree with the argument of Gaver et al. (2004) that probes foster intimacy between designers and users. This method helped us build bridges between us, the architects, and the inhabitants.

5. Conclusions: Design probing as a method for resilience building

Design probing allowed us to produce exercises that duly considered ecological and social aspects. The exercises aimed at fostering social sustainability through people's engagement and the focus on their ways of living. Moreover, the exercises also raised awareness and interest in ecological issues. By directing the participants' attention to aspects of their everyday life and their environment through the probes, the participants may feel encouraged and empowered to tackle these issues themselves. When the inhabitants find that their participation matters and could translate into development, their feeling of empowerment and ownership is enhanced and might encourage further actions to develop the sustainability of their community.

The examples presented in this chapter illustrate the potential design probing has as a technique to support a community's ability to adapt to change and to keep developing without losing its core characteristics. When design probing is geared toward sustainable development and building resilience, both inspiration and information are needed, as well as empathy at the deepest possible level within the constraints of the project. For architects working in developing countries, utilizing a method that supports these qualities can be an asset. Furthermore, our experiments showed that design probing can be less time-consuming and more adapted to the rapid urbanization pace of cities than traditional participatory-design practices. However, as probing is only one possible technique among others, future research would need to consider probing in relation to other empathic-design methods when targeting resilience building.

We can conclude that design probing as a participatory method for resilience building provides a possibility for the inhabitants to take part in the process of change and for architects to learn about the core characteristics of the community and its inhabitants' aspirations and dreams. At this stage of the process in Zanzibar, we cannot yet find long-term evidence that empathic design and probing exercises lead to augmented resilience and sustainability; however, indications in this direction are possible to detect. Our experiences hint at the probing exercises having an impact on community engagement and empowerment. For instance, in his probing responses, Ali had wished for an apartment where he could have a small shop downstairs. As the affordable-housing project was not implemented fast enough, he ended

FIGURE 8.9   The house of Ali's aunt, Ali's shop, and the *baraza* he created in front of it. *Credit: Helena Sandman (2017).*

up transforming a shed attached to his house into a small shop, in front of which he organized a *baraza* around a tree on public land (Fig. 8.9).

Another example is Mama Barke, who runs a small coffee shop in Ng'ambo and participated in our activities. She told us that after our project she established a discussion group together with some of her clients, inhabitants of the neighborhood. They intended to discuss their views on how they would like their neighborhood to develop in the future, a novel experience of public debates for them, according to Mama Barke. The small positive changes demonstrated in the cases of Ali and Mama Barke support the overall hypothesis that participatory engagement enhances resilience.

Acknowledgments

We would like to thank all the participants of the probing exercises, the helpful staff of the Department of Urban and Rural Planning of Zanzibar, as well as our research assistants and translators Saada Wahab and Emma Nkonoki. Additionally, we are grateful for the support of Mariana Rantanen and Ivan Segato in the housing-design process.

References

Akama, Y., & Yee, J. (2016). Seeking stronger plurality: Intimacy and integrity in designing for social innovation. In C. Kung, E. Lam, & Y. Lee (Eds.), *Proceedings of Cumulus Hong Kong 2016* (pp. 173–180). Hong Kong: Hong Kong Design Institute.

Boehner, K., Gaver, W., & Boucher, A. (2012). Probes. In C. Lury, & N. Wakeford (Eds.), *Inventive methods: The happening of the social* (pp. 185–201). London: Routledge.

Correa, C. (1994). Special address: Great city, terrible place. In I. Serageldin, M. A. Cohen, & K. C. Sivaramakrishnan (Eds.), *Proceedings of the second annual World Bank Conference on Environmentally Sustainable Development* (pp. 45–52). Washington, DC: World Bank.

Debrah, R. D., De La Harpe, R., & M'Rithaa, M. K. (2017). Design probes and toolkits for healthcare: Identifying information needs in African communities through service design. *The Design Journal, 20*(1), 2120–2134. https://doi.org/10.1080/14606925.2017.1352730.

Folke, C. (2006). Resilience: The emergence of a perspective for social-ecological systems analyses. *Global Environmental Change, 16*, 253–267. https://doi.org/10.1016/j.gloenvcha.2006.04.002.

Folkers, A. S. (2014). Planning and replanning Ng'ambo – Zanzibar. *South African Journal of Art History, 29*(1), 39–53. Retrieved from https://journals.co.za/content/sajah/29/1/EJC157764.

Gaver, W., Boucher, A., Pennington, S., & Walker, B. (2004). Cultural probes and the value of uncertainty. *Interactions, 11*(5), 53–56. Retrieved from http://research.gold.ac.uk/4720/1/p53-gaver.pdf.

Gaver, W., Dunne, A., & Pacenti, E. (1999). Cultural probes. *Interactions, 6*(1), 21–29. https://doi.org/10.1145/291224.291235.

Hume, D. (1739/1896). A treatise of human nature. Book III, part I, section I. In L. A. Selby-Bigge (Ed.), *A treatise of human nature by David Hume, reprinted from the original edition in three volumes and edited, with an analytical index.* Retrieved from https://oll.libertyfund.org/titles/hume-a-treatise-of-human-nature.

Hussain, S., Sanders, E. B.-N., & Steinert, M. S. (2012). Participatory design with marginalized people in developing countries: Challenges and opportunities experienced in a field study in Cambodia. *International Journal of Design, 6*(2), 91–109.

Juma, M. (2014). *Zanzibar HUL initiative, International Expert Workshop & Stakeholder Conference, Zanzibar.* Retrieved from http://www.historicurbanlandscape.com/themes/196/userfiles/download/2014/6/30/gao1gqnktnahlak.pdf.

Koskinen, I., & Battarbee, K. (2003). Introduction to user experience and empathic design. In I. Koskinen, K. Battarbee, & T. Mattelmäki (Eds.), *Empathic design: User experience in product design* (pp. 37–50). Helsinki: IT Press.

Leonard, D., & Rayport, J. F. (1997). Spark innovation through empathic design. *Harvard Business Review, 75*(6), 102–113.

Maibom, H. (2017). Introduction to philosophy of empathy. In H. Maibom (Ed.), *The Routledge Handbook of Philosophy of Empathy* (pp. 1–10). Oxon: Routledge.

Mattelmäki, T. (2006). *Design probes.* Helsinki: University of Art and Design.

Mattelmäki, T., Vaajakallio, K., & Koskinen, I. (2014). What happened to empathic design? *Massachusetts Institute of Technology Design Issues, 30*, 67–77.

Mazé, R. (2007). *Occupying time: Design, time, and the form of interaction.* Karlskrona: Blekinge Institute of Technology.

Pallasmaa, J. (2015). Empathic and embodied imagination: Intuiting experience and life in architecture. In P. Tidwell (Ed.), *Architecture and empathy* (pp. 4–18). Helsinki: Tapio Wirkkala Rut Bryk Foundation.

Sanders, E. B.-N., & Stappers, P. J. (2008). Co-creation and the new landscapes of design. *CoDesign, 4*(1), 5–18.

Sandman, H., Levänen, J., & Savela, N. (2018). Using empathic design as a tool for urban sustainability in low-resource settings. *Sustainability, 10*(2493). https://doi.org/10.3390/su10072493.

Sandman, H., Maguire, T., & Levänen, J. (2020). *Unboxing empathy: Reflecting on architectural design for maternal health* (Manuscript submitted for publication).

Simonsen, S. H., Biggs, R., Schlüter, M., Schoon, M., Bohensky, E., Cundill, G., … Moberg, F. (2015). *Applying resilience thinking: Seven principles for building resilience in social-ecological systems.* Retrieved from http://stockholmresilience.org/download/18.10119fc11455d3c557d6928/1459560241272/SRC+Applying+Resilience+final.pdf.

Smeenk, W., Tomico, O., & Van Turnhout, K. (2016). A systematic analysis of mixed perspectives in empathic design: Not one perspective encompasses all. *International Journal of Design, 10*(2), 31–48.

Steen, M. (2011). Tensions in human-centred design. *CoDesign, 7*(1), 45–60. https://doi.org/10.1080/15710882.2011.563314.

Stockholm Resilience Center. (2015). *What is resilience?.* Retrieved from http://www.stockholmresilience.org/research/research-news/2015-02-19-what-is-resilience.html.

Stojanovic, T., McNae, H., Tett, P., Potts, T. W., Reis, J., Smith, H. D., & Dillingham, I. (2016). The "social" aspect of social-ecological systems: A critique of analytical frameworks and findings from a multisite study of coastal sustainability. *Ecology and Society, 21*(3), 15. https://doi.org/10.5751/ES-08633-210315.

UNISDR. (2017). *Terminology. Resilience.* Retrieved from https://www.unisdr.org/we/inform/terminology.

United Nations. (2018a). *Sustainable Development Goal 11.* Retrieved from https://sustainabledevelopment.un.org/sdg11.

United Nations. (2018b). *LDCs at a glance.* Retrieved from https://www.un.org/development/desa/dpad/least-developed-country-category/ldcs-at-a-glance.html.

Walker, B., & Salt, D. (2006). *Resilience thinking: Sustaining ecosystems and people in a changing world.* Washington, DC: Island Press.

Part II—Architecture and urban design to enhance community preparedness

Consolidation design as an adaptation strategy in the Toi Market, in Nairobi, Kenya

Georgia Cardosi[1], Susan N. Kibue[2], Mauro Cossu[1]

[1]Faculté de l'Aménagement, Université de Montréal, Montreal, Quebec, Canada; [2]Department of Architecture, Jomo Kenyatta University of Agriculture and Technology, Juja, Nairobi, Kenya

OUTLINE

Enhancing Disaster Preparedness
https://doi.org/10.1016/B978-0-12-819078-4.00009-5

1. Precariousness, uncertainty, and slum consolidation in Nairobi

In Nairobi, the capital of Kenya, about 180 slums and tens of unauthorized markets are called "informal settlements" (SDI Kenya & AMT, 2017). These settlements, where about 70% of the population live and operate (MuST, SDI, UoN, & UCB, 2012), have formed without planning or permission by the authorities, through diverse and often illicit mechanisms of land occupation. The informal arrangements arise because the government cannot meet the urban poor's needs regarding affordable housing and services (Anyamba, 2011). Informal markets sustain the urban economy and provide the population with affordable food and goods (Onyango, Wagah, Omondi, & Obera, 2013). Nevertheless, tenure insecurity constantly threatens the traders, as it hinders infrastructure and service development, and favors eviction and demolition, which are rampant in Nairobi. In 2018, various slums and informal markets (such as Woodley, Adam Arcade Market, Mutindwua Market, Buruburu, and part of Kibera) were demolished by the Nairobi County to implement infrastructure and middle- and high-income housing projects (Oduor, 2018). Some demolitions came after a one-month notice was issued on July 19, 2018 by the Secretary of the National Buildings Inspectorate, instructing the occupants to vacate the land or risk being forcefully removed (Kimuyu, 2018).

Thus, life in unauthorized markets, like in slums in general, can be very uncertain. However, we noted that despite the traders' precarious conditions their design attitude is not as compromised. In analyzing design initiatives in one market, rather than considering slum-related spatial problems from the point of view of design professionals, we recognized some adaptive mechanisms adopted by the traders. Ultimately, this analysis intended to bridge the identified gap between "(social) adaptation and resilience theory" (Norris, Stevens, Pfefferbaum, Wyche, & Pfefferbaum, 2008) and the "design theory" anchored in adaptation (Simon, 1996). Thus, we focused on one of the main aspects of the human–environment relationship: adaptation and settlements' transformations.

This chapter presents the partial results of a doctoral research conducted at the University of Montreal, which explored the relationships between design and adaptive capacities in the Toi Market, Nairobi's second-largest informal market, which has endured an almost 40-year existence. The research recognized three main design approaches at different phases of the market's life: developmental design, evolutionary design, and consolidation design. Whereas developmental design favored the market's formation and the development of its structural features over time, evolutionary design determined its sudden radical socio-spatial transformations. Conversely, consolidation design overlapped with these two approaches. This intrinsic case study (Stake, 1994) helped to understand the actual and potential value of design in the market's formation, stabilization, and development.

Although in Section 5 we provide a brief description of developmental design and evolutionary design, this chapter focuses on consolidation design. This design approach emerged in relative social stability, that is, during the absence of internal social tensions or controversies with the authorities. We elaborated in Section 2 the theoretical framework of design as adaptation, in Section 3 the methodology, and in Section 4 a description of the Toi Market. In Section 5 we present the empirical results and in Section 6 we discuss the role of design in contested urban spaces together with practical and theoretical implications. Finally, in Section 7 we reflect on the limits between adaptation and survival.

2. Design as adaptation: Herbert Simon's approach to reading design in slums

The investigation of design in the Toi Market was initially guided by Herbert Simon's approach to design as adaptation. First, Simon's epistemological premise that "everybody designs who devises courses of action aimed at changing existing situations into preferred ones" (Simon, 1996, p. 111) allowed us to consider slum dwellers as designers. Second, he studied adaptation through behavioral components: decision-making and problem-solving. In the Toi Market, we considered design as decision-making related to its spatial organization.

Contested by constructivist-design theorists and philosophers as a rational, linear logic (Dreyfus, 1972; Huppatz, 2015; Schön, 2005; Soo Meng, 2009), Simon's view of design as human adaptation to the environment offered nonetheless powerful research tools. Simon (1996) contended that the human search for alternatives occurs in conditions of bounded rationality as designers tackle uncertainty about the future and costs in acquiring information that guides decisions. Therefore, humans make decisions by "satisficing" (Simon, 1996, p. 30), that is, choosing the options that are not necessarily the best, but at least achievable. Design, therefore, results from decisions taken under the influence of various environmental factors and fosters adaptive mechanisms.

Several authors describe precariousness and uncertainty as typical conditions in slums (Dovey & King, 2011; Fernandes, 2011; Pamoja Trust, 2008). Environmental decay, poverty cycles, and social unrest constantly affect the lives of the Toi Market's traders. Yet, they have endured and envisioned a future while designing multiple solutions that have contributed to the market's perpetuation. Whereas design and architecture have typically been considered as a sort of artistic inspiration within the professional dimension (Cross, 2006; Lawson, 2009; Rittel & Webber, 1973; Schön, 2005), the literature review indicated that the design skills of slum dwellers and underpinning adaptation mechanisms are poorly understood.

Design as a human natural ability has been considered by authors such as Alexander (1964), Papanek (1972), and Rapoport (2003). Yet, Simon's perspective remains the one that directly addresses adaptation and, therefore, best fits our case study. As Soo Meng (2015) emphasized, Simon's approach to design provides a constructivist opening based on the notion of bounded rationality. This includes, for instance, the idea of "design without goals" (Simon, 1996, p. 162), a bounded-rationality strategy that welcomes the unexpected consequences of design and integrates them as new goals. This and other adaptation mechanisms identified by Simon provide an interesting framework to analyze the design in contested urban spaces. Also, being detached from a disciplinary orthodoxy, Simon remains a reference relevant to today's design studies (Sarasvathy, 2013). Although various authors (Cross, 2007; Lawson, 2004) have recognized that design is an empirical science that requires specific research methods, these have not yet been used and a Simonian ontology of the artificial world has not yet been axiomatized (Forest & Micaëlli, 2009).

The study of design made by slum dwellers contributes to the design theory by exploring its role in adaptation to risk and spatial consolidation. Slum's formation, consolidation, and development should be understood beyond the rationales of informality and disorder. Order and logic in informal settlements' growth are discerned by analyzing decision-making related to spatial organization, and the emanating vision of the future. At the Toi Market, we focused on the thought processes and actions aimed at improving the traders' living conditions through the market's physical transformations. Hence, we analyzed how design thinking and actions differed during the settlement's development periods according to the external conditions.

3. A mixed case-study strategy: Longitudinal and cross-sectional explorations

Our longitudinal case study resorted to cross-sectional investigations and participant observations. Six fieldwork stages were carried out between 2004 and 2016. Participant observations consisted in organizing and attending events; meeting with the traders, community leaders, and market groups; and taking part in activities such as the survey of the market sections in 2007. We shared dozens of experiences with the traders to understand their perceptions of uncertainty related to the insecurity of tenure, and their future motivations and expectations following their spatial-organization actions. We paid attention to the leaders' sociopolitical roles, and to how decision-making was affected by the community's internal dynamics. During the last fieldwork, data collection comprised 59 semistructured interviews, informal conversations with market traders and leaders, and three meetings with the main groups: the market committee, the Toi Upendo Ladies, and Muungano Wa Wanavijiji (the Kenyan Federation of Slum Dwellers).

Longitudinal analysis consisted in observing the market's physical transformations and identifying patterns of spatial arrangements. We mapped the market's layout in three key moments of its history (2007, 2011, and 2016), and produced drawings of stall typologies and photographs. We used *Google Earth* satellite maps from 2001 to 2019 to document the changes in the market's typo-morphology over the years. We supported these observations with the study of 80 relevant documents from the market-committee office, a structure that hosts the committee meetings and keeps stall registers and other documents relevant to the market's management. We used a multiscale approach by investigating design at three levels: (a) the unit of commerce, or individual activity; (b) the market sections or interest groups; and (c) the entire market area or groups of governance.

4. The Toi Market: A contested space

The Toi Market provides for a multitude of goods and services in the Kibera constituency. The market area of about 3.5 ha was originally government-owned land destined for public services. During the 1980s, traders living in the Kibera slum gradually occupied the area, giving life to a large and polyvalent market. In the 1990s, the then City Council of Nairobi subdivided the land and allocated, irregularly, some plots to private entities who eventually claimed them. A legal case is still pending to decide on the legitimacy of these claims and the rights of the traders (Kimotho, 2013). Due to the uncertain tenure regime, the consequent conflicting interests, and the lack of building permits, the market's existence has been continuously threatened by politicians and external groups attracted to this prime land. Over time, the community has thus fragmented into several different groups, ultimately contributing to a socioeconomic and political crisis.

Interviews and documents from the market archives indicated that since the market's inception, the traders have faced five eviction attempts, two complete demolitions, one politico-ethnic conflict, and two fires that razed it down. Lack of security of tenure emerged as the main factor of instability. The traders' living conditions are aggravated by inadequate drainage and sewer systems, potable water, and solid-waste management. Abundant seasonal rains cause floods that affect structures and circulation routes, worsening the already scarce hygienic conditions. Further, fragile structures and poor materials, combined with illegal power connections, augment the vulnerability to fire. These risks create different levels of stress for the traders and exacerbate their precariousness.

5. Adaptation strategies through design

The Toi Market was formed through the initiative of a group of street traders living in Kibera. They lacked formal jobs and searched for a place to settle to sell goods. Thus, they set up their stalls, which multiplied in tandem with the activities' growth. Spatial-organization decision criteria (location and use) followed predetermined group rules, such as the distribution of types of goods. Food kiosks concentrated around the northern access; bananas in the northeastern section; second-hand clothes and shoes on the western side, along one of the main pedestrian circulation routes; the metal-works artisans settled in the eastern section; and cereals along the only internal road in the south.

A basic governance structure comprising local officials—a chairman and a secretary—liaised, on behalf of the traders' then small community, with the officials of the City Council of Nairobi to secure the occupation of this public space. The market densified along existing irregular pathways and by the end of the 1980s, the area counted thousands of temporary stalls. Fig. 9.1 shows the dense and irregular market layout in 2007. This preliminary structure, though in a temporary settlement, allowed the traders to withstand removal notices and demolition threats over the years. Eventually, the efforts put into shaping the nascent market set strong social bonds that encouraged the traders to rebuild it after its first demolition by the State Police, in 1995. The spatial arrangements that emerged in this initial phase can be qualified as developmental design, a stage that marks the passage from a temporary market to a semipermanent one.

During the postelection conflict undergone by Kenya in 2008, external groups set the market on fire. These groups claimed the land ownership, but the market leaders managed to negotiate by including them in the market reconstruction and stall allocation. We characterize this phase as evolutionary design as the traders decided to transform the market's irregular layout to a quasi-formal and rational one based on an orthogonal grid, as shown in Fig. 9.2.

FIGURE 9.1   The Toi Market in 2007. *Credit: Georgia Cardosi (2007).*

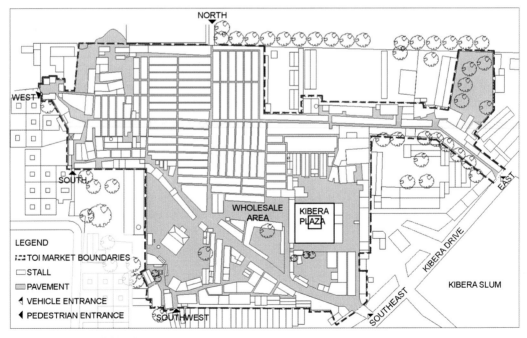

FIGURE 9.2   The Toi Market after 2008. *Credit: Georgia Cardosi (2016).*

The regular urban form improved the visibility and accessibility to all market sections. Plans for the new layout were developed by the market leaders with the aid of a technician. The new layout used planning principles from a redevelopment project that had been devised by the leaders with an NGO before the 2008 crisis.

Although the market layout changed, no infrastructure was put in place. After the 2008 crisis, the market life recovered with difficulty, and risks and threats went on as usual. Indeed, the land-tenure issue is still far from being resolved. In March 2019, a new fire—causes of which remain unclear—destroyed about 80% of the market. A few days later, the traders started rebuilding it.

This brief account illustrates that the market underwent three construction—disaster—reconstruction cycles. Each time, the traders rebuilt the market without governmental permission and minimal financial aid. Concerning the first reconstruction in 1995, we did not get any information about received support. In 2008, the traders relied on their resources and the support of Jamii Bora (a Kenyan bank supporting low-income earners), Pamoja Trust (an NGO that advocates for slum dwellers), and Muungano Wa Wanavijiji for the reconstruction process. In the 2019 reconstruction, the Nairobi County Council provided iron sheets.

We identified 28 types of design initiatives emerging during these cycles: 15 modified the entire settlement or its parts, whereas 13 tackled the stall unit. Through their analysis, we understood the market transformations over the years, its emergence and growth process, including consolidation trajectories. This analysis allowed us to organize the design done by the traders into the three categories or approaches: developmental design, evolutionary design, and consolidation design.

Developmental design consists of collective adaptive initiatives undertaken during periods of social instability and in response to urban interventions such as evictions and demolitions. This type of design aimed at increasing survival opportunities by securing a space where the market's functions could be established. This design course expresses resistance to a hostile environment and willingness to find room for development in the urban context. Evolutionary design instead refers to a macro-transformation in the market's urban form. Manifested as an adaptation to significant external events, it occurred at a specific moment, in 2008, and indelibly altered the market's layout. These definitions of design should not be intended as rigid categories, for they are not exhaustive and their boundaries overlap. Rather, they provide insights on the traders' design capacity and types of design initiatives used in the Toi Market to reduce risks. Table 9.1 outlines the main design practices conducted during the three phases.

5.1 Reducing eviction risk by consolidating the market

Periods of relative stability in the Toi Market's history corresponded to the phases following reconstructions when the community did not have major events to face and simply coped with daily and ordinary stressors. In these phases, the design was carried out at the group and individual levels and entailed alliances among market members. Design initiatives aimed at stabilizing the market's community by trying to secure land tenure and income. The design also created a more coherent physical environment, through the construction of permanent and semipermanent structures and the emergence of spatial patterns.

TABLE 9.1 Examples of design actions undertaken at the Toi Market.

Phase	Level	Design action
Developmental design	Individual and/or group	— Search for a location and settlement in the area — Set-up of vending structures — Incremental aggregation of stalls with the creation of (irregular or regular) sections — Formation of empty areas for public and community uses — Creation of services
Evolutionary design	Group	— Survey of existing activities and spaces before the fire — Numbering of sections and stalls — Plans to distribute the stalls in rows — Plans of the circulation system — Design of the stall module
Consolidation design	Group	— Ablution blocks — Toilet blocks — Group-meeting halls — Drainage line

In Subsections 5.2 and 5.3 we analyze design initiatives during the consolidation phase, either at the group or at the individual level.

5.2 Design at the group level

Consolidation design has been carried out by committees composed of about ten people, which work on construction projects through negotiations, planning, and project management. They have been doing design through meetings, reports, budgets, and plans, supported by external technicians. Interviewees involved in collective projects explained what a project entailed for individuals and groups. According to the analysis of the interviews, the traders' responses included having a sense of belonging, economic survival, development, and common objectives with a vision for the future. In 2011, an interviewee highlighted that *"a project can empower groups to develop a sense of community, bringing people together to shape a future and generate income."* Problem-solving through design enhanced intergroup and intragroup bonds that stabilized social units of various entities. Muungano Wa Wanavijiji, for instance, connects the women and youth groups. Ultimately, despite having different interests, the groups strengthened their ability to advocate for their rights.

Interviewees in 2011 also viewed projects as *"something"* that starts with different people for common interests and that can have a positive impact on both the individual and the community. When geared toward the future implementation of projects serving all groups, design thinking facilitated the development of ethical values for those involved and the improvement of their common ideals for development. Design thinking eventually enabled this twofold community commitment toward the future, through adaptation mechanisms such as prediction and feedback processes. The latter enabled the traders to plan and implement a different market layout in 2008, thus improving accessibility and visibility. Also, some traders envisioned developments or risks, and the vision of the market's future encouraged them to invest in it individually or collectively. These mechanisms allowed for the positive bridging of past experiences and future expectations. In the following subsections, we analyze three projects to illustrate this level of design and its outputs.

5.2.1 Design initiatives under risk of eviction and demolition

In conditions of high uncertainty related to evictions and demolitions, the traders designed accordingly, forecasting these situations. Community leaders designed organizational strategies of land occupation to secure tenure and propose alternatives to eviction and demolition attempts by the authorities. Their design was centered on rights advocacy, which is mainly led by members of the Muungano Wa Wanavijiji. The interviewees mentioned that authorities typically resist demolishing permanent structures, built with concrete bricks because such a move is likely to find opposition and cause demonstrations. Also, permanent structures contribute to upgrading the market.

Recently, leaders of saving schemes also tried to engage the traders to design solutions for the community as a whole. The land-tenure aspect largely impacts on some traders' vision of the market's future. With the land case pending, the traders have been forecasting different spatial configurations adapted to probable results. As some project-committee members informed us, political and legal discourses have been converging with physical design solutions. Although the directly concerned traders are those having stalls under the contested plots, other traders proposed that if the land case is lost, the affected activities should be redistributed within the market. They asserted that no trader should be excluded from the market and lose their livelihood. In case the claimants lose and the government repossesses the land, they planned to negotiate on how they would buy the land as a community. Some interviewees further suggested negotiating a few hectares of land to keep the market working while allowing for the government to implement the public services originally planned by the Nairobi City Council.

Some traders reasoned about the need, if the case is lost, to replace the affected services and businesses, an alternative that sounds difficult, if not impossible. As Fig. 9.3 shows, the contested area includes the market's mosque, the wholesale and garbage areas, three toilet blocks, two garage areas, and part of the cereals section (stalls along the main road at the south), besides many other activities established in the southern area. These spaces host key functions and losing them would affect the market's operations. Moreover, the land case has implications for the market's boundaries, which, in turn, impact the development of infrastructures. Until the market's land and boundaries remain undefined, and the market is not legalized, the county government cannot start any planning process to provide these.

To devise well-defined spatial solutions, the traders' design thinking goes beyond negotiations with the authorities. The traders who engage in spatial-organization initiatives explored ways to include all traders on the land, by optimizing the available space. For instance, they proposed the reduction of stall dimensions and the opening of new footpaths in some sections. Some members of the Muungano Wa Wanavijiji also proposed the stalls' modernization, with their relocation into separate departments.

If the traders win the case, the government may repossess the land and the risk of market demolition by third parties would decline. Then, they might negotiate with the government to gain land ownership. The community might try to establish a cooperative to raise funds for

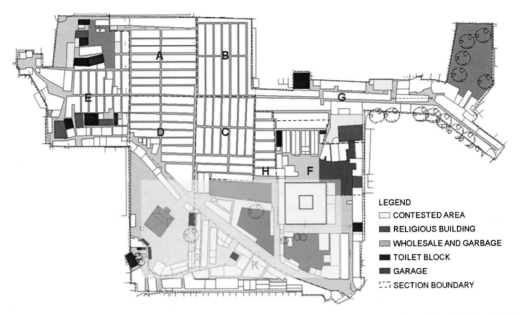

LEGEND
☐ CONTESTED AREA
▨ RELIGIOUS BUILDING
▨ WHOLESALE AND GARBAGE
■ TOILET BLOCK
▨ GARAGE
⌐⌐ SECTION BOUNDARY

FIGURE 9.3   Parts of the contested land in the Toi Market in 2019. *Credit: Georgia Cardosi and Patrizia Piras (2019).*

the purchase, whereby a savings scheme would help every member to contribute daily (for example KES10) to collect KES40 million (USD400,000) over a three- to four-year period. Based on their experience and without the support of third parties, some interviewees suggested creating a members' registry and a survey of all traders to guarantee their inclusion and implement a communal form of land property.

As an alternative, some interviewees focused on the need to upgrade the market without having land ownership. They also suggested that the Nairobi County Council could contribute to the upgrading as it collects revenue from the traders. This type of design thinking is driven by the awareness of common objectives related to survival and the recognition of the traders' rights. It bears witness and contributes to the *sense of belonging* in the community. It was noted, however, that not all the traders wanted to consolidate and develop the market; some even opposed such moves. Interests largely differ among market groups and individuals and this variation leads to a high fragmentation within the market's contested land.

5.2.2 Design initiatives under environmental risks: The drainage project

The Toi Market has always lacked a proper drainage system. Existing drainage lines fall under the supervision of Muungano Wa Wanavijiji, which typically undertakes negotiations with the Nairobi City County. This system presents open-air and underground channels that lack connections and prove insufficient in the rainy season. Most of the market sections are affected by flooding during heavy rains. In 2016, when a small donation was available for projects, most interviewees identified drainage as a major problem in the market, especially in Section A (see Fig. 9.3), where many food kiosks are located on both sides of open rainwater drainage channels. Improving this drainage line, which would benefit the entire market, was also identified as a priority by leaders and participants at meetings of the Muungano savings scheme and the Toi Upendo Ladies.

The traders initiated a drainage-improvement project in Section A, after benefitting in October 2016 from a workshop on sanitation conducted by one of the market's members, Mr. Geoffrey Okwaro, then a Master's student of Civic and Development Education at the Catholic University in Nairobi. Following this event, the traders selected a 17-member committee, including interest groups (by gender, ethnicity, and trade type), and sections' representatives. Their main priority was the improvement of the drainage channel, an 80-m long line with open and underground segments.

The committee subsequently worked on an affordable budget and raised contributions from representatives of the food kiosks' area. They prepared and published tender documents for the project, to which two contractors responded. The construction started in December 2016 but was unfinished due to insufficient funds. The committee mobilized all market members to contribute and ultimately raised additional funds to complete the project. They further reached out to the Nairobi County assembly to solicit for financial support and to the National Youth Service to provide building materials for the open drains. However, none of these contributions materialized and the works ground to a halt. Fig. 9.4 shows the initial and incomplete works.

FIGURE 9.4 Initial works to install the underground drainage channel. *Credit: Geoffrey Okwaro (2017).*

Owing to the deterioration of the environmental conditions there was an imminent health and eviction problem. Indeed, the authorities could order the market to close due to the prevailing unsanitary conditions, as had previously occurred in 1999. Consequently, the project was not granted the necessary formal building permission, as informal communities without tenure security remain excluded from the urban-planning realm.

5.2.3 Reducing environmental risks to avoid eviction: The Muungano Wa Wanavijiji ablution block

The Muungano Wa Wanavijiji ablution-block project consisted of a set of toilets and showers for men and women, designed and built between 2012 and 2016. The project was supported by Pamoja Trust, the Nairobi Water Company, and Athi Water Service Board, and funded by the European Union.

In the Toi Market, the Muungano federation had previously built another toilet block in 2007. Up to then, toilets were few because of high construction costs and the lack of permits to connect to the public sewer. Toilet blocks proved to be very strategic projects, not only for reducing environmental and eviction risks but also for increasing income opportunities. Firstly, they derive income from the payments made by the users, and secondly, they provide youth with employment opportunities. Moreover, the provision of clean toilets fostered a better shopping experience and increased the market's customer base, resulting in the traders seeking to multiply them.

Despite its benefits, the 2012 Muungano project met several challenges. Construction dragged on from 2012 to 2016 and the project created conflict even before it began. As the former market chairman explained, this toilet block exemplified six types of challenges common to projects in slums. The first one was defining the location: looking for an adequate project site that was convenient to all traders was a difficult process. Moreover, this project needed space that did not belong to any group. To implement the toilet block, the traders first identified an old pit latrine, at a space managed by the youth group. Thereafter, they bought, through funds from the Muungano saving scheme, a vacant land claimed by the youth group. The second challenge was competition and envy among members and groups. The third one was taking into account the ongoing situation on the ground. For instance, the 2012 toilet project could not connect to the existing sewer lines as its drainage invert levels were not aligned to those of the existing sewer line, thus making it necessary to construct another one.

The fourth challenge related to traders' attitudes and involvement in the projects. For instance, a Muungano member noted that traders tend to believe that projects—either toilets or others—are inevitably income-generating activities. Referring to projects funded by donors, he argued that *"community groups often revert to a form of 'dependency syndrome', expecting profit from the project without effort."* This dependency from donors' funds *"makes that various projects in Kibera supported by NGOs collapse when donors withdraw."* This posed a challenge to the sustainability of donor-funded projects. The social transformations that come with such projects take time to occur but are eventually achieved through sound management and consultations guided by clearly outlined project objectives. A problem identified in the toilet project was that many Kibera residents still use flying toilets—plastic bags for human-waste disposal—due to lack of toilet facilities. This attitude can be mitigated by incorporating into the project's objectives a vision aiming at altering social behavior.

The fifth challenge common to projects in slums is corruption: in this project four contractors quitted without paying the laborers, leaving the works unfinished. Indeed, without a stipulated agreement, contractors can mismanage the funds and delay the project. According to some interviewees, *"one of the contractors was a politician who used the money for a political campaign."* Finally, there is the challenge of security, as the doors and some sanitary fittings were stolen, a problem not anticipated by the project committee. Hence, two guards were deployed, with an increase in the project's costs.

In 2016 the project was not yet working well due to competition with other toilet blocks—there were 14 permanent or semipermanent toilet blocks in the market, belonging to either groups or individuals. Clients and traders tended not to use the new toilets, a fact that was aggravated by a lack of information about their functioning. Finally, the project resulted in poor financial revenue to the community.

5.3 Design at the stall level

By the occupation and transformation of the surrounding spaces, traders improved their stalls to increase their businesses. These changes could only take place following an agreement and collaboration between the members sharing the contiguous spaces. The individuals' and alliances' spatial-organization initiatives ultimately generated spontaneous design patterns, illustrated in the following subsections. The design of the stalls included changes in their structural parts, such as roofs, partitions, doors, and windows (Table 9.2). Such changes included improving the environmental conditions within and around the stalls, beautifying the accesses and protecting them from weather conditions, and increasing spaces for stock.

5.3.1 Raising and extension of roofs

Roof raising presents various advantages. In food kiosks, for instance, this solution improves internal ventilation (Fig. 9.5). In second-hand clothes stalls, this helps to enhance sales by creating more space to display the products, which increase the stocks' capacities and

TABLE 9.2 Categories of traders' design actions in consolidation design.

#	Individual actions undertaken at the stall level
1	Raising and extension of roofs
	— To catch up light and ventilation
	— To increase visibility
	— To protect customers and goods
2	Combining stall modules
3	Increasing internal comfort and beautifying solutions
4	Creation of verandas
5	Adaptation to open-air drainage line
6	Occupation of public spaces
7	Construction of second floors

FIGURE 9.5   Roof raising in a food kiosk. *Credit: Geoffrey Okwaro (2018).*

hence the businesses. The additional income enabled the traders to join saving schemes, generate capital, and invest in other stalls. Roof extensions help to protect goods from sun and rain and reduce the losses of perishable products. While roof extensions have become a pattern, only the wealthy can afford this solution.

5.3.2 Combining stall modules

The stalls in the Toi Market consist of modular units of two basic sizes: 2×2 m and 3×2 m, with a structure made typically with eucalyptus poles, roofed with iron sheets, and enclosed in fabric and polythene sacks. These units are organized in rectangular rows and mirrored on their longer side to form sections. When the traders can afford to own more units, they often combine them to create larger trading spaces. Unit combination occurs for stalls facing each other along a secondary passage, generating an intermediary private– public zone, or mirroring them on the back wall. In Fig. 9.6, for instance, a tailor eliminated the central partition wall, doubling space and openings on the pathways. This design strategy helps to attract clients.

5.3.3 Increasing internal comfort and beautifying solutions

When resources became available, the traders tried to personalize their stalls and improve their comfort and aesthetics to benefit their clients. Despite being temporary structures, their stalls were arranged to accommodate elements such as chairs and changing rooms (clothes stalls), and the goods were displayed in an attractive manner (Fig. 9.7).

5.3.4 Creation of verandas

Many stalls have front verandas protecting entrances from rain and sun. While roof exten- sions typically occur over pathways and do not use vertical-support elements, verandas appear as an integral *architectural* part of the units. For instance, the veranda of the Toi Mar- ket's Muungano office works as a branding feature of the institution (Fig. 9.8). Stronger veranda roofs can also provide additional space for the installation of equipment.

FIGURE 9.6   Second-hand clothes stall doubled to have two accesses. *Credit: Georgia Cardosi (2019).*

5.3.5 Adaptation to open-air drainage lines

Some existing ground conditions also required adaptive solutions. For example, the traders selling cereals along the main access road needed a solution for the open drainage channel located there, which generally created a barrier for the structure to extend toward the pathways. The traders were able to turn a presumed problem into a design solution, which consisted in arranging a raised platform placed over the drainage channel to display cereals. This arrangement, shown in Fig. 9.9, has also become a pattern.

5.3.6 Occupation of public spaces

It was noted that the internal space of stalls in some less-crowded sections was often empty. This depended on the lack of stocks and/or on the traders' attempts to optimize the opportunities to display the merchandise at the front. Therefore, in periods of stability, the occupation of public pathways was evident through these displays' design patterns. Typically, a semipublic realm emerges at the stall front, where the trader and customers interact. These transitional/transaction zones are vital to the trade. Fig. 9.10 shows the corresponding spatial sequence. The transitional/transaction zones (in yellow) overlap with both the internal private space (in gray) and the external public passage (in red).

FIGURE 9.7   Stall for the sale of household items. *Credit: Georgia Cardosi (2019).*

FIGURE 9.8   The Toi Market's Muungano Wa Wanavijiji office. *Credit: Patrizia Piras (2011).*

The traders also tend to cover public passages with polythene sacks. This practice to protect people and goods from the sun and rain has become another design pattern. Despite their ephemeral character, these sacks introduce a colored atmosphere in the market space. Roof extensions instead acquire a more stable aspect and look like *aerial* expansions of the stalls, as if the traders ambitioned to put their mark on the skyline (Fig. 9.11). At times some traders invented sophisticated solutions beyond the market's standards. For example, the owner of two stalls separated by a pathway built an arched-like roof to connect them. Despite being a beautifying solution, this initiative embodies the risk of shutting off access to the wholesale area by this owner (Fig. 9.12).

5.3.7 Construction of second floors

Structural extensions with a second floor are another common design initiative and bear witness to major investments in periods of stability. These extensions might be used to secure

FIGURE 9.9   Design solution to adapt to drainage lines. *Credit: Patrizia Piras (2016).*

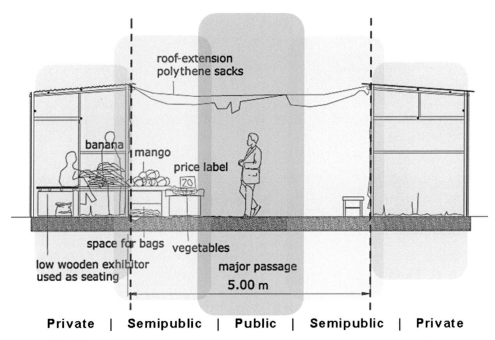

FIGURE 9.10   Spatial sequence in a market passageway. *Credit: Georgia Cardosi (2016).*

land tenure but might also be for generating additional income, whereas some structure owners in less profitable sections have started turning stalls into residential spaces for rent. Such extensions might come at the expense of tenants, who might one day be evicted. Structural extensions to include residential use may eventually threaten the market's existence.

FIGURE 9.11   A particularly exuberant roof extension. *Credit: Patrizia Piras (2019).*

FIGURE 9.12   An arched-like roof over a public passage. *Credit: Georgia Cardosi (2019).*

6. Discussion: The role of design in managing vulnerability

One often questions the role of design and designers in today's trajectories of urban development, when vulnerability factors rise due to rapid societal changes. If the value of professional design in offering adequate solutions in ever-growing urban complexity has been long questioned (Rittel & Webber, 1973; Schön, 2005), the value of nonprofessional design has been scarcely recognized. Concerning informal settlements, it is still difficult to explain the existence of design and its forms. Our study shows that even the vulnerable context of the Toi Market presents intense and well-rounded design activities carried out by the traders. This fact makes us reflect on what we call the *slum paradox*: Precariousness, uncertainty, and vulnerability do not hinder slums' formation, proliferation, and consolidation. It is within the dimensions of such a paradox that we identify the role played by nonprofessional designers and their adaptive mechanisms for survival and existence under contemporary urban challenges.

At the Toi Market, we learned that design initiatives do not diminish with precariousness and uncertainty; rather, they get more robust and acquire specific directions. Therefore, the design is not only a prerogative of favorable conditions. Also, design as adaptation cannot be attributed to separate disciplines or sectors as its expressions converge to assure survival through spatial, economic, and social considerations. In the market, it was evident that some traders were aware of design priorities and solutions. Learning from nonprofessional design adaptations revealed gaps in traditional professional-oriented design thinking in uncontested spaces. Indeed, in the latter, the emphasis on spatial considerations may undervalue other important although less tangible design requirements. Further, the research indicated that daily, ordinary, and exceptional stressors made a difference in the way design was conceived and produced in the market. Accordingly, we identified three approaches: developmental, evolutionary, and consolidation design.

Consolidation design as adaptation manifested in a twofold manner: strategic and tactical. On one hand, design initiatives at the group level targeted social cohesion and rights advocacy. On the other, design at the individual level tended instead not to use centralized planning attempts, such as those that guided the market reconstruction in 2008—understood here as evolutionary design.

An outcome of trying to manage demolition and eviction risks through consolidation was evident in design initiatives aimed at densifying the market's fabric. Yet, such compactness increased the potential for fire risk. A second outcome was the forecasting component that built upon a vision for the future. Overall, consolidation design seemed to work on risk reduction in the short term, rather than having a transformational power. The community has been withstanding cycles of construction—disaster—reconstruction in a quasi-continuum mode but could not avoid internal social disruptions. High vulnerability was also nurtured by internal scissions caused by social and ethnic tensions.

Finally, Simon's premise mentioned in Section 2 might be reformulated to fit informal settlements as "everyone in contested urban spaces designs who devises courses of action aimed at reducing vulnerability and risk exposure." At the same time, shifting from the paradigm of "preferred conditions" to "needed survival conditions" urges one to research unexplored design approaches. However, addressing design as decision-making and an adaptation process in contested urban spaces opens many questions. For instance, the extent to which design contributes to managing urban risks is yet to be explored. Similarly, although we considered design as a potential attitude of every person, not everyone is a designer. Further research is needed to illustrate users' design skills in contested urban spaces and systematize the knowledge about design-problem representation and design tools. Viewed this way, a systematic study of nonprofessional design can largely contribute to Simon's yet unfinished design theory (Hatchuel, 2002).

7. Conclusions: What future for the Toi Market?

The Toi Market's traders have been enduring difficult conditions, yet their spatial-improvement interventions are noticeable. These *developments* should nonetheless not be understood as an upgrading of the traders' working conditions, which continue to be affected by scarcity of resources and lack of access to infrastructure, services, and tenure. Furthermore, the study poses new questions and doubts on the market's future and the role of design in it, as existing precarious conditions continue to undermine its existence.

The arson event of March 2019 engendered a new disaster—reconstruction cycle that raises critical questions. Precariousness of tenure indeed hinders a proper development intended as a shared improvement of environmental and socioeconomic conditions. Such precariousness eventually reinforces the traders' adaptation capacity and, consequently, design abilities. We are cognizant of the fact that the traders' design adaptation mechanisms are need driven and they often have limited choices in their actions. What we discuss as adaptation is hence more a survival issue than one of fitness and development.

After the March 2019 fire, the traders rebuilt the market in a few days to restart business and food distribution, thus manifesting design abilities in the rapidity of the physical recovery and the capitalization upon previous construction experiences. Yet, adaptation still does not equate to improved living conditions, and the vulnerability cycle is not interrupted. How can the traders get out of this first stage of adaptation toward paths of development? Is there any effective long-term development in the robustness of this cyclical mechanism of failures and recoveries? How can one recognize it through design's adaptive mechanisms? We pose these questions for future research.

Although acknowledging the design abilities of nonprofessionals and their practice as a tool for social cohesion in contested spaces, we did not intend to deny the responsibilities of governments and civil society organizations in the development trajectories of vulnerable populations. Similarly, we did not intend to provide simplistic ways to deal with the Toi Market's land-tenure issue, a delicate legal and political imbroglio related to the occupation and use of a public land. Instead, we hope that unveiling the traders' vision of the future embedded in their design practices can help the stakeholders interested in the market land to envision ways to negotiate solutions that fit most needs in a satisficing and comprehensive way.

Acknowledgments

We thank the Toi Market community for welcoming us along this research. Our special thanks go to Patrizia Piras, a former member of the Architecture Open Circuit NGO, who volunteered during the fieldwork and drew most of the Toi Market's maps. We also recognize the contributions for the 2016 fieldwork by Nelson Mwangi, Donald Kiplangat, and Hakim Bisher, students of the School of Architecture and Building Sciences of the Jomo Kenyatta University of Agriculture and Technology (JKUAT), and Grace Kariuki, an alumnus of JKUAT. Finally, we recognize the support of the Disaster Resilience and Sustainable Reconstruction Research Alliance (Œuvre Durable) and the Observatoire Ivanhoé Cambridge du Développement Urbain et Immobilier.

References

Alexander, C. (1964). *Notes on the synthesis of form*. Cambridge, MA: Harvard University Press.
Anyamba, T. (2011). Informal urbanism in Nairobi. *Built Environment, 37*(1), 57—77. https://doi.org/10.2148/benv.37.1.57.
Cross, N. (2006). *Designerly ways of knowing*. London: Springer.
Cross, N. (2007). From a design science to a design discipline: Understanding designerly ways of knowing and thinking. In R. Michael (Ed.), *Design research now: Essays and selected projects* (pp. 41—54). Basel: Birkhäuser.

Dovey, K., & King, R. (2011). Forms of informality: Morphology and visibility of informal settlements. *Built Environment, 37,* 11–29. https://doi.org/10.2148/benv.37.1.11.

Dreyfus, H. L. (1972). *What computers can't do.* New York, NY: Harper & Row.

Fernandes, E. (2011). *Regularization of informal settlements in Latin America.* Cambridge, MA: Lincoln Institute of Land Policy.

Forest, J., & Micaëlli, J.-P. (2009). *A Simonian ontology of the artificial world.* Paper presented at the CONFERE '09, Marrakech, Morocco. Retrieved from https://halshs.archives-ouvertes.fr/halshs-00539642v2/file/A_simonian_ontology_of_the_artificial_world.pdf.

Hatchuel, A. (2002). Towards design theory and expandable rationality: The unfinished program of Herbert Simon. *Journal of Management & Governance, 5*(3), 260–273.

Huppatz, D. J. (2015). Revisiting Herbert Simon's "science of design." *Design Issues, 31*(2), 29–40. https://doi.org/10.1162/DESI_a_00320.

Kimotho, R. (June 30, 2013). *Kenya's contribution to the UNDP report legal empowerment strategies. Draft, 2* (Unpublished Manuscript).

Kimuyu, H. (August 8, 2018). Illegal structures at Mutindwa demolished: More in city estates targeted. *Nairobi News.* Retrieved from https://nairobinews.nation.co.ke/news/illegal-structures-mutindwa-demolished.

Lawson, B. (2004). *What designers know.* London: Routledge. https://doi.org/10.4324/9780080481722.

Lawson, B. (2009). *Design expertise.* Oxford: Architectural Press.

MuST, SDI, UoN, & UCB. (2012). *Mathare Zonal Plan | Nairobi, Kenya. Collaborative plan for informal settlement upgrading.* Retrieved from http://knowyourcity.info/wp-content/uploads/2015/04/Mathare_Zonal_Plan_25_06_2012_low_res-2.pdf.

Norris, F. H., Stevens, S. P., Pfefferbaum, B., Wyche, K. F., & Pfefferbaum, R. L. (2008). Community resilience as a metaphor, theory, set of capacities, and strategy for disaster readiness. *American Journal of Community Psychology, 41*(1–2), 127–150. https://doi.org/10.1007/S10464-007-9156-6.

Oduor, B. (July 26, 2018). *What the law says about demolitions.* Daily Nation. Retrieved from https://www.nation.co.ke/lifestyle/dn2/What-law-says-about-demolitions/957860-4681054-kj4cehz/index.html.

Onyango, G. M., Wagah, G. G., Omondi, L. A., & Obera, B. O. (2013). *Market places: Experiences from Kisumu city.* Kisumu, Kenya: Maseno University Press.

Pamoja Trust. (2008). *An inventory of the slums in Nairobi.* Retrieved from https://knowyourcity.info/wp-content/uploads/2015/04/Nairobi_slum_inventory_jan_09.pdf.

Papanek, V. J. (1972). *Design for the real world: Human ecology and social change.* New York, NY: Pantheon Books.

Rapoport, A. (2003). *Culture, architecture et design [Culture, architecture, and design]* (S. E. Sayegh, Trans.). Gollion: InFolio (in French).

Rittel, H. W. J., & Webber, M. M. (1973). Dilemmas in a general theory of planning. *Policy Sciences, 4*(2), 155–169. https://doi.org/10.1007/bf01405730.

Sarasvathy, S. D. (2013). MAZES without minotaurs: Herbert Simon and the sciences of the artificial. *European Management Journal, 31*(1), 82–87. https://doi.org/10.1016/j.emj.2012.11.002.

Schön, D. A. (2005). *The reflective practitioner: How professionals think in action.* Aldershot: Ashgate.

SDI Kenya, & AMT. (2017). *An integrated development plan: Mukuru SPA inception report* (Technical paper). Retrieved from https://www.muungano.net/s/Mukuru-SPA-Inception-Report.pdf.

Simon, H. A. (1996). *The sciences of the artificial* (3rd ed.). Cambridge, MA: MIT Press.

Soo Meng, J. C. (2009). Donald Schön, Herbert Simon and *The Sciences of the Artificial. Design Studies, 30*(1), 60–68. https://doi.org/10.1016/j.destud.2008.09.001.

Soo Meng, J. C. (2015). Design without final goals: Getting around our bounded rationality. *Artifact, 3*(4), 2.1–2.7.

Stake, E. R. (1994). Case studies. In K. N. Denzin, & S. Y. Lincoln (Eds.), *Handbook of qualitative research* (pp. 236–247). Thousand Oaks, CA: Sage.

10

Risk and urban design in Brazilian favelas: Linking participation, collective spaces, and territorial management

Pablo Benetti, Solange Carvalho

Post-Graduation Program in Urbanism, Faculty of Architecture and Urbanism, Federal
University of Rio de Janeiro, Rio de Janeiro, Brazil

1. Introduction

The absence of a housing policy in Brazil offering affordable urban land suitable for low-cost housing means that low-income populations, excluded from the logic of the formal market and the protection of the State, are obliged to resort to irregular land occupation and the informal urban land market, following the "logic of necessity" (Abramo, 2003).

Enhancing Disaster Preparedness
https://doi.org/10.1016/B978-0-12-819078-4.00010-1

Ownership tends to be uncertain in these lands, leading to their gradual and haphazard occupation. Contrary to what occurs in market- or State-driven urban expansions, in which infrastructure regularly precedes occupation (Solà-Morales, 1997), in those cases the latter comes before the former. In other words, there is no proper urbanization of the sites before their occupation with buildings. This is how favelas usually emerge, developing in urban areas that are not necessarily appropriate for residential use and that are often subject to environmental risks. Most favelas in Rio de Janeiro and other major Brazilian cities are located on hills and wetlands, being prone to flooding and landslides caused by heavy tropical rains. Therefore, geomorphological and hydrometeorological risks are an intrinsic problem of these settlements. Very often, favelas' territories include areas designated as risky that need to be vacated or require technical solutions.

In the 1990s, Brazil began implementing slum-upgrading policies to mitigate urban conflicts and risks. These policies have gained worldwide prominence for their integrated intervention approach (Fiori, Riley, & Ramirez, 2001; Magalhães & Villarosa, 2012). Such modality is defined by an emphasis on "multidimensional, multisectoral, multiprogrammatic, and multiscalar [urban] projects that understood the requalification of site as inherently related to the requalification of its multiple articulations with context and ultimately with the city" (Fiori, 2014, p. 44). These programs were initially focused on urban improvements and infrastructure, but risk mitigation gradually became one of their main guidelines, in accordance with international development agendas.

The best-known slum-upgrading program in the world (Magalhães & Villarosa, 2012), Rio de Janeiro's Favela-Bairro (1994—2007), had a major focus on improving urbanity, but the related projects also had to point out the existing risks and address them. Gradually, risk management has become an issue of greater importance and priority within these initiatives. Thus, more specific risk-management programs have been implemented all over Brazil since the late 2000s, such as PAT-PROSANEAR (Low-Income Sanitation Technical Assistance Project) or PAC/PPI/Integrated Sanitation (Growth Acceleration Program/Priority Investment Project). These federal programs focused on sanitation, mitigation of environmental risks, and resettlement of populations subject to socioenvironmental hazards through an integrated intervention approach, just like in the slum-upgrading policies.

In the city of São Bernardo do Campo (metropolitan region of São Paulo), for example, the "elimination of risk situations" became a guideline in favelas' upgrading (Regino, 2017, p. 72), to be achieved either through urban-design solutions or by removing families from risk and resettlement areas to safer ones. Yet, urban risks in favelas are a very delicate issue, which can be used for political interests, as it was the case in Rio de Janeiro, following the landslide disasters caused by heavy rains in February 2009. The Mayor Eduardo Paes was then accused of using the risk-mitigation argument to remove families from favelas to resettle them in less favorable areas in the city (Heck, 2013). In this sense, and to avoid speculative actions, spatial projects continue to be a powerful tool to guide risk-mitigation public actions as a driver of transformations in these urban settlements (UN-Habitat, 2017), even though these policies have been facing setbacks in Brazil following spending cuts.

We understand that risk mitigation, urban design, resilience, and urban management are strongly connected. Solutions to mitigate risk should strive to include local inhabitants in their proposals, for they can defend the resulting spaces as theirs and rebuild them in a resilient manner when necessary. Such a framework means a new way of conducting urban management in favelas, not by the State alone, but together with the local inhabitants. In this context, we conceive resilience as the transformation capacity that seeks to improve the system (Vale, 2012), rather than maintaining a preexisting, deficient, and unsustainable urban status, to guarantee the safety and permanence of the inhabitants in the upgraded favela.

In this chapter, we present and discuss these aspects using concrete experiences of urban-design projects in the favelas of Rio de Janeiro and São Paulo, in which we observed how the reoccupation of hazard-prone areas continues to be a recurrent phenomenon even after public interventions. These dynamics indicate a permanent friction between urban design and the management of the favela's territory, two actions carried out by the State.

This chapter starts by discussing the tension between risk mitigation and urban design in the favelas, presenting issues related to urban informality and the role the project plays in the processes of urban improvement and disaster risk reduction (DRR). The following section contains the cases of urbanized slums, organized by the themes of risk mitigation, urban management, and participation. Finally, some of the studied cases pinpointed that neither urban design nor the management of these territories, when really carried out, recognizes the internal logic of the favelas, a fact that threatens the sustainability of DRR initiatives. Other cases showed that the involvement of local actors in projects and in partnerships with the State contributes to DRR, but that without the effective presence of the State in these territories DRR efforts cannot be sustainable.

2. Risk mitigation and urban design in favelas

The demand for housing often makes urban risks seem like a distant, lesser problem. Living in a risk area means being exposed to hazards that can, someday, materialize into disasters. Yet, not having a place to live means being exposed on a permanent basis to social risks. For instance, the absence of a stable residence implies an invisibility that cuts off access to social registries for education, health, and employment. In this sense, having no place to live is almost synonymous with not formally existing. To choose between living in a risk area and being exposed to social risks is a very tense and cruel decision. Moreover, the cost of urban land is so high that leaving empty even hazard-prone areas seems a wastage of resources. This fact is even more acute in favelas that are located close to employment opportunities, public transportation, and urban facilities. These dimensions of social vulnerability should not be overlooked in risk-mitigation policies (Marchezini & Wisner, 2017).

The strong connections between risks and social vulnerability are discussed by several authors. For instance, as pointed out by Beck (1992), socio-environmental inequalities and the location of poverty go hand in hand. "The history of risk distribution shows that, like wealth, risks adhere to the class pattern, only inversely: wealth accumulates at the top, risks at the bottom" (Beck, 1992, p. 35). The poor often live in precarious places not because of some risk preference, but because of the lack of other possibilities. Based on this fact, Bullard (1994) proposed the concept of "sacrifice zones," identifying urban spaces where the evils of predatory industrial development—such as atmospheric and noise pollution, or other imminent risks—are concentrated, as illustrated for instance by neighborhoods around a nuclear power plant. In connection with the Brazilian context, Cunha et al. (2015) further developed this idea, considering favelas to be outstanding "sacrifice zones," that is, segregated territories where the highest environmental burdens are concentrated in discriminated and low-income populations.

In the same sense, pointing out the relation between risks and poverty, Thouret (2007, p. 89) stated that the poverty "forces people to live in less expensive but dangerous areas, and dominates the daily concerns of people who do not have the means—in terms of money and time—to preserve the environment." In an extreme statement, we could say that environmental risks in favelas would only be eliminated when social, physical, and institutional vulnerabilities no longer existed. Public interventions such as those reported in this chapter are aimed at reducing vulnerabilities in such a way that potential risks are mitigated, thus reducing the likelihood of disasters and losses of lives and property.

In urban-design projects, risk mitigation encompasses multidisciplinary activities, such as solid-waste collection, drainage solutions, building contention barriers, reforestation, cleaning rivers and streams, developing roads, in addition to raising the awareness of the residents regarding the risk of occupying unsuitable areas. In favelas, risk mitigation often involves the removal of housing from hazard-prone areas and the need to make the resulting spaces useful to avoid reoccupation. This is one of the greatest difficulties of urban projects in favelas. Although the array of possible DRR solutions is quite diverse, their estimated costs remain a key factor in choosing the most suitable ones. Depending on the estimated costs, risk mitigation may go beyond the budget allocated to a slum-upgrading project, thus having to be provided by another dedicated government agency or program that will complement the urban improvements. In the worst case, costs can even make the upgrading of the favela unfeasible.

This is the type of situation in which urban-design projects can play a key role, as they can foster investments to be as efficient as possible. Connecting the demands of governmental programs with local needs in the same intervention becomes a project strategy that enables the implementation of priorities—such as risk mitigation and the development of a new public open space—as an integrated solution. The question is how to stop a vicious circle of initial urban occupation of risk areas, their clearance (pursued as an initiative for both DRR and the provision of public spaces), and their subsequent reoccupation when the intervention is concluded. In other words, how can one intervene in risk areas in such a way that they do not become again a threat to local residents?

DRR solutions often include the creation of open areas (normally called "public spaces") in hazard-prone residential areas. The cities' public spaces have been the object of governmental interventions of urban administrations since the 1980s. The term *public spaces* yet gives the illusion of democratic spaces of "peaceful and harmonious coexistence with the heterogeneousness of society" (Delgado, 2011, p. 20), while in fact concealing a policy of exploitation, control, and segregation resulting from capitalism. We thus adopt the term *collective spaces*, understood to be free-access spaces that can foster relationships with the public power and certain local social bonds. In slum-upgrading interventions, even if the idea of providing open spaces as a DRR measure is guided by good intentions, in some cases these areas are reappropriated for private uses. These situations suggest that the concept of democratic space is not really applicable in favelas, where it is perhaps better to think of these spaces as belonging to a group—namely a part of its inhabitants—and not to the public in general.

That condition signals a sharp difference between formal areas and favelas. In the former, the State usually guarantees the continuity of the public nature of the collective spaces using urban codes as a support. In favelas, the State leaves when the urban-design intervention is concluded, and the lack of written urban codes that force the State to be present creates an unregulated situation in which parallel powers can establish their own rules. Therefore, the idea of collective spaces varies greatly between favelas and other parts of the city (Benetti, 2017). Collective spaces in the formal city are supported by a combination of rules, sanctions, and established behaviors that preserves the limits between private and public areas, even if collective spaces can support some private uses (such as tables on the sidewalks, informal vendors, etc.) without being properly privatized. Private limits are generally preserved in favelas, but such is not the case of collective spaces.

The history of these territories shows a progressive encroachment on open areas to accommodate private necessities. The pressure to occupy more spaces—for necessity or for speculation—is always present in favelas. Spaces without a clear destination are sources of permanent tension and potential conflict. This tension comes from both the conflict inherent to the formation of favelas, where empty spaces can mean the possibility of expanding one's home or for a lot to be speculated on (Benetti, 2017; Carvalho, 2009), and the top-down project methodology imposed by public programs carried out without the involvement of residents in the setting of priorities (Roy, 2005).

Interviewed on his experience in the Favela-Bairro Program, Sérgio Magalhães, former Secretary of Housing of the Municipality of Rio de Janeiro (1993–2000), recalled *"the potentiality of the project as an articulator of public policies—of urbanization and project design—to enable the coordination of the various policies that were previously isolated"* (personal communication, 2018). However, despite this important role in upgrading favelas, the project is often not able to sustain its actions. The lack of continuity in post-construction urban management is a factor that compromises the sustainability of public actions in favelas (Magalhães & Villarosa, 2012).

In this regard, Fiori (2014, p. 47) warned that we still need to "develop spatial strategies that engage very explicitly with politics of institutional transformation" to articulate "spatial design with the re-design of urban-political institutions and regulations." Certainly, such an institutional redesign should begin with the required legal and social recognition of favelas as places with a different culture of construction. This shift would imply adapting the public institutions responsible for oversight and maintenance to favelas' reality, in which the commitment of the residents seems to be a sine qua non prerequisite for the integrated management of these territories.

3. Some cases of risk mitigation and urban design in favelas

In this section, we present examples of urban design in favelas in which the State tried to address urban risks (see Table 10.1). Based on the considerations above, we selected cases in which DRR measures have been carried out by government-funded slum-upgrading programs considered complete by their respective managers. We also reviewed these programs' methodologies and the transformations actually imprinted in the territory. We assessed how these transformations were achieved and how urban design contributed to them. None of the cases presented is yet recognized by the literature as best practices. We believe that it is necessary to learn not only from best practices, which constitute models for the international planning nowadays (Roy, 2005), but also by looking at what went wrong in less-successful initiatives, to achieve more efficient urban-design methodologies for sustainability and resilience.

TABLE 10.1 General data related to the cases presented in the chapter.

Favela	Area	Dwellings	Inhabitants	Municipality	Slum-upgrading program	Sponsor	Intervention type
Babilônia	85,956 m^2	1178	2451	Rio de Janeiro	Bairrinho Program (2002–2004)	Municipality and European Union	Slum-upgrading project
					Morar Carioca Program (2011–2015)	Municipality and Federal Government (PAC)	
Cantinho do Céu	Project (the new park): 380,000 m^2 Favela: 1,548,761 m^2	8016	28,042	São Paulo	Mananciais Program—Billings Reservoir (2008–2018)	Municipality and Federal Government (PAC)	Part of the slum-upgrading plan of favelas around the Billings Reservoir
Chapéu Mangueira	34,595 m^2	401	1288	Rio de Janeiro	Bairrinho Program (2002–2004)	Municipality and European Union	Slum-upgrading project
					Morar Carioca Program (2011–2015)	Municipality and Federal Government (PAC)	
Grotinho, in Paraisópolis	Project: 4200 m^2 (open spaces); 1000 m^2 (buildings) Favela: 822,739 m^2	17,141	45,694	São Paulo	Slum-upgrading Program (Second phase: 2008–2010)	Municipality and Federal Government (PAC)	Part of the slum-upgrading plan of Paraisópolis
Parque São Bernardo	—	3020	—	São Bernardo do Campo	Integrated Slum-upgrading Program (2009–2017)	Municipality and Federal Government (PAC)	Slum-upgrading project
Sítio Bom Jesus	47,470 m^2	749	2741	São Bernardo do Campo	PAT-PROSANEAR Program (2008–2018)	Municipality and Federal Government (PAC)	Slum-upgrading project

We conducted field visits in the cities of São Paulo and São Bernardo do Campo in 2016 and 2017, whereas the cases in the city of Rio de Janeiro are part of both authors' more than 20 years of experience with urban-design projects in favelas. We also interviewed architects, urban authorities, and favelas' residents to understand the initial idea and post-work evolution. Our objective was to look at the previous situation (when the area was considered more risky) and the subsequent one (after the mitigation of the related risk) and verify how the project, the execution, and the ensuing maintenance operations were carried out. We were ultimately guided by the desire to identify situations unsuccessfully managed through design, that is, cases that challenged the professional practice of architects and urban designers showing limitations in risk-mitigation initiatives in urban projects in favelas.

This analysis was based on partial outcomes of Carvalho's doctoral research, which aimed at understanding the challenges, limitations, and results of urban design in favelas. The objective of her Ph.D. research was to contribute to improving the performance of these projects, and to increasing the effectiveness of slum-upgrading policies and the efficiency of related investments (Carvalho, 2020).

The urban-design projects and the cases of urban management that we present tried to propose solutions for risk areas and sought, in different ways, to deal with the specificities of favelas. These cases portray some of the solutions adopted to mitigate urban risk in favelas. They show how difficult it is to preserve the intervention as initially proposed and how new places often emerge subsequently. They also show how the logic of permanent spatial dynamics in favelas introduces several challenges to managing and sustaining these new places after the commissioning of the related structures. Therein lie our proposal and contributions. The issues we tried to cover are related to the following question: how can risk mitigation in favelas be effective and sustainable over time? This query led us to discuss both the project and its management.

3.1 Examples of risk-mitigation initiatives

In Parque São Bernardo (city of São Bernardo do Campo), after a disaster that led to the collapse of houses, the risk of landslides was managed with the construction of a retaining wall at the bottom of the slope. As this intervention was assumed to be only a technical solution, it did not involve urban design to also attach a meaningful purpose to it. Unfortunately, this solution did not stop the reoccupation of the slope. The wall in the upper part of the slope has been partially demolished and the earth behind it now constitutes an informal landfill that has been developed by the residents after the conclusion of the DRR works. As informed during a field visit by the municipality officer in charge of the area, residents started to create a "stable" ground using bags filled with rubble to provide parking spaces (as shown in Fig. 10.1). Residents of the houses located below the retaining wall have complained that the newly created ground is putting the downhill houses at risk again.

In the case of Sítio Bom Jesus (also in the city of São Bernardo do Campo), in an area prone to floods, the riverbanks were vacated by integrating collective spaces used as circulation and leisure areas, landscaping, and risk-mitigation measures. The residents were involved in the project decisions to address this risk area and, to date, there has been no reoccupation of the resulting spaces (Fig. 10.2). This case shows that it is possible to transition away from the top-down logic, in which ideas and proposals put forward by the residents are often minimized

FIGURE 10.1   In Parque São Bernardo, the white bags on the slope have been used to carry rubble to recreate land. *Credit: Solange Carvalho (2017).*

or ignored by the staff of public agencies, almost like the relationship between the "oppressed" and the "oppressors" (Freire, 1970/2005). The methodology of the Sítio Bom Jesus project, which placed the residents as protagonists in the design's decisions and discussions related to the various risk-mitigation options, seems to be an important reason why its apparent results have been maintained over time. Yet, it is important not to confuse project participation with information or communication about the design. The effective participation of the project's users is an instrument to foster the endurance of the shared use of the collective spaces designed in favelas. In this regard, Benetti (2017, p. 101) noted that

> the project cannot simply be a formal act. Its sustainability and permanence depend on the social bonds established during its elaboration and the subsequent implementation works. There is no guarantee that [collective spaces] will remain [collective spaces] in the future and that the agreements that made [them] possible will be maintained over time. Only the consolidated social use can guarantee the permanence of the project, when each resident who participated in its elaboration […] is, in a sense, an ally in defending and preserving [these collective spaces].

The example of Grotinho, part of the favela of Paraisópolis in São Paulo, shows the fragility of collective spaces (Fig. 10.3). To prevent deadly disasters due to landslides,

FIGURE 10.2   In Sítio Bom Jesus, collective spaces were implemented to prevent the reoccupation of flood-prone water banks. *Credit: Solange Carvalho (2014).*

FIGURE 10.3   In Grotinho, the transformation of the collective space resulting from the integrated DRR intervention: as inaugurated in 2009 (left) and in 2017 (right). *Credit: Fábio Knoll (2009) and Solange Carvalho (2017).*

Grotinho benefited from public investments to vacate the slope, to remove more than 200 families from the risk area, and to construct collective spaces and a multipurpose building for public use. The objective of the intervention was to give a shared use to the new urban area that resulted from the houses' removal and thus prevent its reoccupation. As shown in Fig. 10.3, this action was not enough. The building, the hill, and the square are in the process of being occupied by drug traffickers, who expanded their activities by subdividing the land eight years after the completion of the project's construction. The square at the bottom of the hill, now completely destroyed and covered with garbage, has become unrecognizable.

Ensuring the preservation of open spaces that have been designated for collective use and preventing their reoccupation and the ensuing reemergence of urban risks depend on the social appropriation that can be structured around them. It is not enough to design and build certain solutions. However, in the Grotinho case, there was no agreement between local forces and the State to take care of these areas. Armed criminal agents who control the territory have shown that they have more power than the population, which, abandoned by the State in the hands of a parallel authority, has been left with no recourse.

In addition to physical interventions, it is key to develop alliances between the residents and the potential users of each place, who can then become more inclined to defend the collective use of these areas against unwanted private occupations in the future. Project methodologies must either incorporate these issues or be held hostage to ever-recurring formal gestures with very limited sustainability. This, however, does not invalidate the need for the presence of the State to regain control of these territories, often under the grip of parastatal groups, which is essential to supporting such collective uses and to consolidating the public sphere in these areas.

3.2 Urban management for preventing risk resurgence

When urban-design and DRR works in a favela are completed, the State usually withdraws, delivering back the modified territory to the local agents—residents, local groups, and residents' associations. These are then forced to coexist with armed militia and/or drug gangs who impose their asymmetric power. In addition, individual spatial initiatives in favelas compromise the new spaces created through urban-design projects. These two processes lead to the likely resurgence of urban risks in these territories. Moreover, favelas are seldom effectively incorporated into the management of the city. Thus urban oversight and public maintenance are absent or only occur intermittently. As identified by Magalhães and Villarosa (2012), we can affirm, based on the cases reviewed in this chapter, that favelas continue not to be subject to the same type and level of urban regulations as the rest of the city.

Risk management often remains a priority even after slum-upgrading interventions, due to the horizontal expansion of favelas toward unoccupied areas and the housing densification through vertical expansion. These two expansion forms, typical of favelas' history, are driven by residents' individual efforts, although today this is not necessarily linked to a real need— speculation by small urban entrepreneurs also plays a role in this regard and cannot be overlooked. Slum-upgrading projects frequently give residents the feeling that eviction no longer looms on the horizon and that the area is consolidated, thus resulting in the increase of land value and the expansion of the favela (Abramo, 2003; Cavallieri, 2003). This fact shows that urban management after upgrading is even more influential in ensuring the sustainability of the attained improvements.

As favelas are sometimes located in environmentally sensitive zones, their horizontal expansion often exerts pressure over designated environmental protection areas. For instance, Cantinho do Céu has grown by occupying the banks of a water basin in the city of São Paulo. In this case, the involved threat was groundwater contamination, an environmental risk that although neither apparent nor evident is absolutely real. Water quality depends on the presence of basic sanitation in this favela and on the existence of a green buffer around the banks of the water basin. The slum-upgrading project led by the Municipality of São Paulo rehabilitated the favela and removed houses from the water edges to create a park.

The Cantinho do Céu Park and the urbanization of the favela are considered a reference of urban-design quality and slum-upgrading solutions by the municipal authority (França & Barda, 2012), which has been managing the area since its inauguration, even though the project has been only partially implemented. The presence of the State in the stretch where the park was built has contributed to the sustainability of the project (Fig. 10.4), although some related initiatives have been discontinued, such as the outdoor cinema.

FIGURE 10.4   In Cantinho do Céu, the new park prevents the contamination of the water basin. *Credit: Solange Carvalho (2017).*

However, during the same field visit to the Cantinho do Céu Park, we identified that in another stretch, where houses had been removed but the civil works were interrupted before the park's implementation, reoccupation has already started (Fig. 10.5). Therefore, the efforts that went into registering residents to receive compensations have been lost. The lack of continuity in the park's works threatens the success and the sustainability of the public interventions in Cantinho do Céu.

In Rio de Janeiro, the horizontal expansion of favelas frequently occurs in local forests, compromising soil stability as the vegetation on the hills helps to prevent landslides. Therefore, the preservation of forests and reforestation are important risk-prevention activities. One of the first attempts to control the horizontal expansion of favelas' territory was to implement eco-fences—a demarcation made of 1.5-m high metal posts and steel cables—in the 1990s. However, it became evident that factors beyond such spatial demarcation played a role in controlling favelas' expansion. The case of the adjacent favelas Babilônia and Chapéu Mangueira, wedged between the forest and the neighborhood of Leme, illustrates the challenges of using such eco-fences (Fig. 10.6).

Although both emerged in the 1910s, these favelas followed different expansion processes that resulted in contrasting urban morphologies. After the implementation of eco-fences by the State, the residents' associations of both favelas worked to raise awareness among residents to respect these boundaries, with divergent outcomes. On one hand, Chapéu Mangueira is practically still within the boundaries of the eco-fences, but has few open spaces and almost no vegetation, and the built environment is dense. In this case, the action of the residents' association prevented the horizontal expansion of the favela into the forest, but not the housing densification through vertical expansion. On the other hand, the failed attempt to control expansion is clearly visible in Babilônia, where there are now many houses beyond the eco-fence. Babilônia is also more spread out and features vegetation and open spaces.

Controlling the typical processes of horizontal and vertical expansions of favelas after upgrading works is a factor that helps to mitigate risks. We present two examples of control attempts in this regard: the Office for Urban and Social Orientation (POUSO—Posto de Orientação Urbanística e Social) and the reforestation task force in the favela of Babilônia.

FIGURE 10.5   In Cantinho do Céu, the banks of the water basin are being reoccupied. *Credit: Solange Carvalho (2017).*

FIGURE 10.6   Aerial photo of Babilônia (on the left) and Chapéu Mangueira (on the right). *Credit: Tales Lobosco (2011).*

POUSO was created by the Municipality of Rio de Janeiro in 1997, to control and ensure the sustainability of slum-upgrading projects. Composed of an architect or an engineer and social workers, each POUSO local team has the mission to supervise the upgraded favela and provide guidance to residents for new works in their houses. Unfortunately, this policy has never been a governmental priority and today lacks an institutional prominence. One of the "indicators" adopted by POUSO to monitor and control favelas' expansions was to watch for the delivery of construction materials and their destination by doing periodical inspections. Official figures revealed that the favelas benefiting from POUSO interventions have grown more slowly than other favelas in the city (Vial & Cavallieri, 2009), but the presence of POUSO was not enough to halt expansion. In reality, this initiative was a structured policy against the dominant construction logic in favelas. As it was not based on a new urban pact, POUSO was deemed more of a repressive character than a mediation action. This perception is even more relevant when considering that mediation is a key feature of social and spatial relationships in favelas, in which residents' associations play an important role.

Whereas POUSO partially failed because it was viewed by the residents' associations as an external power, successful examples of forest protection have been achieved by the Reforestation Community Program. Led by the Environmental Secretariat of the Municipality of Rio de Janeiro since 1995, this program has been deploying teams of favela residents to manage their territories. The residents are remunerated by the Municipality for their efforts and receive the tools needed to reforest and take care of the Atlantic Forest. In Babilônia, this program evolved into the establishment of a local cooperative, CoopBabilônia, which has been serving for 20 years as an example of a fruitful public—private partnership. Together with the Municipality and Shopping Rio Sul (a commercial establishment adjacent to the community), CoopBabilônia has been funding reforestation activities and eco-tourism initiatives (Fig. 10.7).

In Babilônia, we identified a contradictory process, whereby the successful reforestation task force with local labor coexists with the favela's horizontal expansion. The process of new houses being built in the forest has become more pronounced since 2015 (Fig. 10.8),

FIGURE 10.7 Orthophotographs dated 1984 (left) and 2013 (right) showing the reforestation by CoopBabilônia on the slopes next to Chapéu Mangueira. *Credit: Used with permission from Instituto Pereira Passos, Municipality of Rio de Janeiro.*

when the State abandoned the site leaving partially completed the works of the Morar Carioca Program, led by the Municipality of Rio de Janeiro in these favelas (2011–2015).

At the same time, the reforestation program in Babilônia has been contributing to stabilize the soil and hence to prevent landslides. The conservation of the green areas is stimulated and controlled by the members of the cooperative themselves, by promoting social activities like collective planting, eco-paths, and eco-tourism. In this initiative, CoopBabilônia has been a key agent. Today, it preserves the forest and also provides seedlings and reforestation consultancy to other favelas and areas of the city. The interesting feature of this case is the agreement established between the cooperative, inhabitants, local associations, private actors, and the government, leading to an innovative model for urban management that includes different social forces.

FIGURE 10.8 In Babilônia, new constructions in an environmental protection area, initiated after the municipal staff of POUSO left the favela. *Credit: Solange Carvalho (2015).*

3.3 Participation and risk mitigation

The abovementioned examples of government-led projects developed in favelas show the fragility of managing the implemented DRR solutions, be them collective spaces or even public buildings. It is still a big challenge for urban design in slum-upgrading projects to find efficient mechanisms to sustainably address risk issues and adopt solutions that are economically feasible and help to prevent the reoccupation and privatization of open areas, transforming them into democratic spaces. As such, the construction of this alliance between the State and inhabitants must undertake a consistent effort to empower the populations as political subjects who are able to shape the urban-design program from the beginning to the end of the works and beyond. However, this rarely happens and participation is not enough, limited to consulting the populations, who do not have any decision-making power in spite of the pertinent opinions and suggestions they may have. In addition, the participation usually began when the projects' decisions are already taken, and not at the initial stage of the process when budget and concepts are defined. Furthermore, it is not enough to reduce risks by simply removing the populations; it is necessary to endow the newly unoccupied sites with a new meaning.

The issues raised in the cases presented suggest the need to understand the dynamics of the territory that architects and urban designers deal with, and to redesign the urban political institutions and regulations that insist on treating urban improvements in favelas as a charity instead of a right. Favelas carry the marks of their origin and development, whose predominant logic is that of individual efforts with moments of solidarity between neighbors. The culturally rooted occupation of these territories can only be changed if a new relationship pact is created between favelas and the State, ending the current framework in which the State accepts the existence of favelas yet does not accord them permanent legal status. Indeed, even after the end of this type of intervention, the legal status of favelas does not change to integrate them in the formal city. The studied favelas remain "informal" or "illegal," as the regularization process was not implemented immediately after their upgrading and may not even happen, and they continue in a state of permanent transience (Rolnik, 2016).

Slum-upgrading processes are structured in three separate but interrelated moments— project, works, and subsequent maintenance. Although we argue that the same DRR objectives should underlie these three moments, our field visits showed that this rarely happens. As discussed, during the implementation of urban projects in favelas, the State temporarily takes on the management of the space, but normally exits at the end of the works without establishing a strong system to manage the new territory that includes the local forces. This is a fact that occurs not only in favelas but in the city as a whole, as illustrated by the abandonment of the Olympic facilities in the city of Rio de Janeiro after the Rio 2016 Games. Yet, in favelas, where urban management is more fragile, this reality is even more visible, with State investments being a rare occurrence whereas they should be efficient and durable. However, this kind of initiative may possibly result in disappointment, such as in the case of the high investments made in the Complexo do Alemão's cable-car transport system, in Rio de Janeiro. Conceived as a mobility solution for the population, the system has been gradually abandoned since 2016, five years after its inauguration (Ferreira, 2017). Some interventions may also lead to revolt, like in the case of Babilônia (Fig. 10.8), whose residents' association, without any governmental support, found itself unable to stave off construction in environmental protection areas.

4. Conclusions

Based on the studied cases, we conclude that the establishment of a strong management system for DRR interventions in favelas begins with the way in which the urban project is developed. This implies understanding the specific construction logic of favelas, characterized by the lack of written rules and associated with the predominance of individual behaviors. As such, the challenge facing urban-design projects is not only to physically change places, but rather to consider how one can reverse the scenario of prevailing individual interests to that of reinforcing the communities' recognition of urban collective spaces as important components of their living territories. Successful initiatives, such as the one of Sítio Bom Jesus or the reforestation in Babilônia, illustrate that effective urban management is not possible without directly involving the population. At the same time, public management plays a key role in the sustainability of urban improvements. In the cases in which the State "abandoned" the project, the effects were negative, such as the resumption of Babilônia's expansion through the forest or the reoccupation of Grotinho and Cantinho do Céu.

After 30 years of public investments in favelas, residents who have experienced government-led slum-upgrading interventions in their neighborhoods are discussing the importance of their roles as protagonists in these processes. Yet, this is exactly the opposite of what has been practiced, as explained by a community leader: "favela-upgrading projects are conducted in total submission to the contractors [...]. Nobody listens to the population, they only pretend to listen and act according to their own interests" (Pinheiro, 2017, p. 19). The projects' methodologies play an important role in the sustainability of the interventions, identifying the actors who actually imprint their logics on the territory, to ensure their own future survival and that of the project. To achieve successful proposals, architects, urban designers, and other professionals, besides the State institutions that work in favelas, need to understand how different and specific the internal logic of favelas is, as well as their histories, struggles, and resistances. Hence, more adequate methodological approaches for urban-design interventions in these territories should consider all these aspects.

The analyzed cases show that the residents' participation in the definition of the project, local speculative agents, and the management of the spaces resulting from DRR-related urban-development interventions have a crucial role in the subsequent reoccupation of these areas—that is, in the success of risk-mitigation outputs themselves. We believe that understanding these issues can contribute to the consolidation of urban design in favelas as a tool for integrating the management of collective spaces and promoting resilience, serving as an effective instrument for sustainable urban development and risk mitigation.

Acknowledgments

Part of Carvalho's research was funded by the Brazilian agency CAPES (Coordenação de Aperfeiçoamento de Pessoal de Nível Superior), through the PDSE Grant Holder # 88881.189053/2018-01 and the Finance Code 001.

References

Abramo, P. (2003). La teoría económica de la favela: Cuatro notas sobre la localización residencial de los pobres y el mercado inmobiliario informal [Favela's economic theory: Four notes on the residential location of the poor and the informal real-estate market]. *Ciudad y Territorio, Estudios Territoriales, 136*, 273−294. https://doi.org/10.22296/2317-1529.2007v9n2p25 (in Spanish).

Beck, U. (1992). *Risk society: Towards a new modernity*. London: Sage.

Benetti, P. (2017). Costuras urbanas na avenida central do Morro do Alemão: "Isto não é uma praça." [Urban sewing in the central avenue of Morro do Alemão: "This is not a square."]. In P. Benetti, & S. Carvalho (Eds.), *Praça pr'Alemão ter: O germinar de uma praça verde no Morro do Alemão* (pp. 88−101). Rio de Janeiro: UFRJ-PROURB-FAU/Instituto Raízes em Movimento (in Portuguese).

Bullard, R. D. (1994). Urban infrastructure: Social, environmental, and health risks to African Americans. In I. L. Livingston (Ed.), *Handbook of black American health: The mosaic of conditions, issues, policies and prospects* (pp. 315−330). Westport, CT: Greenwood.

Carvalho, S. (2009). Favelas en Rio de Janeiro, Brasil: Interferencias del proceso de urbanismo informal en la vivienda [Favelas in Rio de Janeiro, Brazil: Interferences of the urban-informality process in housing]. In S. G. Padilla (Ed.), *Urbanismo informal* (pp. 223−241). México: Universidad Autónoma Metropolitana−Azcapotzalco (in Spanish).

Carvalho, S. (2020). *Entre a ideia e o resultado: O papel do projeto no processo de urbanização de uma favela [Between idea and results: The role of urban design in the slum upgrading process]* (Doctoral Thesis, PROURB/FAU/UFRJ, Rio de Janeiro, Brazil) (in Portuguese). Retrieved from http://www.minerva.ufrj.br.

Cavallieri, F. (2003). Favela-Bairro: Integração de áreas informais no Rio de Janeiro [Favela-Bairro: Integration of informal areas in Rio de Janeiro]. In P. Abramo (Ed.), *A cidade da informalidade: O desafio das cidades latino-americanas* (pp. 265−296). Rio de Janeiro: Sette Letras/FAPERJ (in Portuguese).

Cunha, M. B., Porto, M. F. S., Pivetta, F., Zancan, L., Francisco, M. S., Pinheiro, A. B., ... Calazans, R. (2015). O desastre no cotidiano da favela: Reflexões a partir de três casos no Rio de Janeiro [The disaster in the daily life of the favela: Reflections from three cases in Rio de Janeiro]. *O Social em Questão, 33*, 95−122 (in Portuguese).

Delgado, M. (2011). El espacio público como ideología *[The public space as an ideology]*. Madrid: Catarata (in Spanish).

Ferreira, P. E. (2017). *O filé e a sobra: As favelas no caminho do capital imobiliário [The steak and the leftover: The favelas in the way of the real-estate capital].* (Doctoral Thesis, FAUUSP, São Paulo, Brazil) (in Portuguese). https://doi.org/10.11606/T.16.2018.tde-22062017-162710.

Fiori, J. (2014). Informal city: Design as political engagement. In T. Verebes (Ed.), *Master-planning the adaptable city: Computational urbanism in the twenty-first century* (pp. 40−47). New York, NY: Routledge.

Fiori, J., Riley, E., & Ramirez, R. (2001). Physical upgrading and social integration in Rio de Janeiro: The case of Favela Bairro. *disP − The Planning Review, 37*(147), 48−60. https://doi.org/10.1080/02513625.2001.10556788.

França, E., & Barda, M. (2012). *Entre o céu e a água: O Cantinho do Céu/Sky and water, the living in between: The Cantinho do Céu* (A. A. Puntch, Trans.). São Paulo: SEHAB-HABI.

Freire, P. (1970/2005). *Pedagogy of the oppressed* (M. B. Ramos, Trans.). New York, NY: Continuum.

Heck, C. (October 29, 2013). The 'area of risk' justification for favela removals: The case of Santa Marta. *RioOnWatch* [Blog]. Retrieved from http://www.rioonwatch.org/?p=11410.

Magalhães, F., & Villarosa, F. (Eds.). (2012). *Slum upgrading: Lessons learned from Brazil*. Washington, DC: IADB.

Marchezini, V., & Wisner, B. (2017). Challenges for vulnerability reduction in Brazil: Insights from the PAR framework. In V. Marchezini, B. Wisner, L. R. Londe, & S. M. Saito (Eds.), *Reduction of vulnerability to disasters: From knowledge to action* (pp. 57−96). São Carlos: RiMa.

Pinheiro, A. B. (2017). Genocídio do favelado [Genocide of the slum dweller]. In A. Magalhães, A. Rossbach, L. C. S. M. Domingues, & P. Benetti (Eds.), *UrbFavelas: Registros e reflexões do II Seminário Nacional sobre Urbanização de Favelas* (p. 19). São Paulo: Publisher Brasil (in Portuguese).

Regino, T. (2017). *Direito à moradia, intervenção em favelas e deslocamento involuntário de famílias: Conflitos e desafios para as políticas públicas [Right to housing, intervention in favelas, and involuntary displacement of families: Conflicts and challenges for public policies]* (Master Dissertation, UFABC, São Bernardo do Campo, Brazil) (in Portuguese). Retrieved from biblioteca.ufabc.edu.br/index.php?codigo_sophia=110484.

Rolnik, R. (2016). Guerra dos lugares: A colonização da terra e da moradia na era das finanças *[War of places: The colonization of land and housing in the financial era]*. São Paulo: Boitempo (in Portuguese).

Roy, A. (2005). Urban informality: Toward an epistemology of planning. *Journal of the American Planning Association, 71*(2), 147−158. https://doi.org/10.1080/01944360508976689.

Solà-Morales, M. (1997). Las formas de crecimiento urbano [*The forms of urban growth*]. Barcelona: UPC (in Spanish).

Thouret, J.-C. (2007). Avaliação, prevenção e gestão dos riscos naturais nas cidades da América Latina [Evaluation, prevention, and management of natural risks in Latin American cities]. In Y. Veyret (Org.), *Os riscos: O homem como agressor e vítima do meio ambiente* (D. Ferreira, Trans.) (pp. 83—112). São Paulo: Contexto (in Portuguese).

UN-Habitat. (2017). *New Urban Agenda*. Retrieved from http://habitat3.org.

Vale, L. (2012). Interrogating urban resilience. In T. Haas (Ed.), *Sustainable urbanism and beyond: Rethinking cities for the future* (pp. 22—23). New York, NY: Rizzoli.

Vial, A., & Cavallieri, F. (2009). O efeito da presença governamental sobre a expansão horizontal das favelas do Rio de Janeiro: Os Pouso's e o Programa Favela-Bairro [The effect of the governmental presence on the horizontal expansion of the favelas of Rio de Janeiro: The Offices for Urban and Social Orientation and the Favela-Bairro Program]. *Coleção Estudos Cariocas, 9*, 1—9. Retrieved from http://portalgeo.rio.rj.gov.br/estudoscariocas/.

PART III

Housing vulnerability, community resilience, and inclusive governance

11

Informality versus short-term regularization of the Syrian refugees' situation in Lebanon

Faten Kikano

Faculté de l'Aménagement, Université de Montréal, Montreal, Quebec, Canada

1. The Syrian refugee crisis in Lebanon

Since 2011, the Syrian conflict has displaced millions of people who now represent the largest refugee population in the world. Most of them sought shelter in neighboring countries. At the beginning of the crisis, Lebanon kept its borders open, welcoming, outside camps, Syrians fleeing the war in large numbers. The country was praised by the international community for its honorable response (Loveless, 2013; Onishi, 2013). However, nearly nine years after the conflict started, Lebanon hosts around 1.5 million Syrian refugees, holding the highest percentage (26.2) of refugees per capita worldwide (UNHCR, 2016).

In addition, it is estimated that Lebanon hosts some 30,000 Palestinian refugees from Syria, 35,000 Lebanese returnees, and a preexisting population of more than 450,000 Palestinian refugees who have been living in 12 refugee camps for almost 70 years (Government of Lebanon & United Nations, 2019). On top of the impact of hosting large numbers of refugees, Lebanon has been weakened by years of war and internal conflict that only ended in 1990, and has been suffering ever since from political instability, insecurity, and corruption, as well as deteriorated services and infrastructure (Fawaz, Saghiyeh, & Nammour, 2014).

This condition has made it harder for refugee populations to assimilate and thrive in Lebanon. As the economic, political, and social situation has worsened and infrastructure and services became more strained, tensions and resentment have risen between host and refugee populations (World Bank, 2013). Moreover, as the Government has not permitted the construction of new refugee camps, refugees have scattered informally all over the country, without any planning or organization (Fawaz et al., 2014). As a result, the urban landscape of the areas with larger refugee populations is being drastically degraded (Dahi, 2014).

In contrast to Jordan and Turkey, whose governments implemented a refugee displacement program, the Lebanese Government delegated the management of refugees to municipalities. These municipalities adopted specific approaches that ranged from hospitality to exclusion, based on the unique needs of each locale (Fawaz et al., 2014). In 2014, overwhelmed with the large numbers of refugees, the Lebanese Government tried to take control of the situation and adopted restrictive measures aimed at preventing new refugees from entering the country and scaling down existing numbers. As in other countries (Crépeau, Nakache, & Atak, 2007), these measures did not reduce refugees' numbers; in fact, they only made refugees and their hosting communities more vulnerable and easier to exploit (Janmyr, 2016). Based on my research into the situation of Syrian refugees in Lebanon, I propose a new policy for hosting refugees in the country.

The chapter is divided into four sections, starting with a presentation of the context in Section 1 and an explanation of the adopted methodology in Section 2. These are followed by an examination of the reasons for, and consequences of, the lack of assimilation of refugees into Lebanon's institutional, socioeconomic, and urban systems in Section 3. In Section 4, the core of the chapter, I propose a series of policy recommendations for the refugee situation. I then conclude with a summary of the main ideas in the chapter.

2. Methodology

Data were collected through six focus-group discussions, three with refugees and three with Lebanese host communities. These discussions took place in localities with large refugee populations: Bar Elias and Kab Elias (in the Bekaa governorate), and Nabaa' (in Greater Beirut). Discussions were undertaken in 2014, 2015, and 2017, which allowed me to evaluate the change in the situation over this period. Semistructured interviews were also conducted in 2017 with three UNHCR officers, four Lebanese ministers, and six municipal representatives. The following themes were examined: (1) refugees' legal status and their freedom of movement; (2) refugees' and host communities' financial situation, availability of jobs, and work conditions; (3) housing availability for refugees and for vulnerable Lebanese, housing types and conditions, number of people versus the size of the dwelling, type of agreement with landlords, and risk of eviction; (4) accessibility of humanitarian funds, and conditions for receiving aid; (5) social acceptance and intercommunal tensions; and (6) perceived level of security.

Information was supplemented with the study of documents and reports from the UN, the UNHCR, the World Bank, international NGOs, and the Lebanese Government. The research focused particularly on the *Lebanon Crisis Response Plan*, a joint initiative of the Lebanese Government and its international partners, notably the EU, the World Bank, and the UN, in line with the commitments made at the *Supporting Syria 2016* conference, during which world leaders agreed to raise funds for Syria and the region (Government of Lebanon & United Nations, 2019).

3. The Lebanese policy: Reasons and consequences

Lebanon's initial response to the Syrian refugee crisis was to maintain open borders with Syria and prohibit the creation of organized camps. According to official sources, the Government established this policy because it did not anticipate that so many refugees would enter the country and it expected the crisis to end soon. Yet, while some scholars describe it as the "policy with no policy" (El Mufti, 2014), other political analysts claim that it was intended to alter the demographic and sectarian balance in Lebanon by increasing the Sunni population in the country, while reducing their numbers in Syria (Minority Rights Group International, 2019). These analysts also maintain that the non-encampment policy had institutional, political, and economic motives as well (Turner, 2015). After 2014, the Lebanese Government adopted restrictive regulations excluding Syrians from legal and formal systems to reduce their numbers in the country. To explain Lebanon's response to the Syrian refugee crisis, it is critical to understand the country's historical approach to managing refugees. I begin by examining the Lebanese legislation regarding refugees.

3.1 Refugee law in Lebanon

Lebanon did not sign the 1951 Refugee Convention and lacks meaningful legislation concerning refugees in its body of law (Doraï, 2006). It was only in 2003, following the Iraqi refugee crisis, that the Lebanese Government signed a memorandum of understanding with UNHCR, wherein Lebanon is defined as a country of transit, not as one of asylum, where refugees are allowed to stay only for a period of one year (UNHCR, 2011).

Consequently, Syrian refugees do not have any legal status in Lebanon. When they register with UNHCR, they are entitled to receive aid, but continue to be subject to the risks of detention and expulsion (Centre Libanais des Droits Humains, 2013). In fact, in official documents produced by the Government, Syrians are "displaced people" (*nazihoun*), a status connotating impermanence (Janmyr, 2016), and without any legal meaning (International Labour Organization, 2014; Naufal, 2012). As for refugee settlements, regardless of their type, they are called as "gatherings" (*tajammouat*) instead of "camps" or "informal settlements" (Government of Lebanon & United Nations, 2019).

The lack of formal laws governing refugees in Lebanon and the fear of camps in the country have a historical basis, present in many countries in the Middle East: the nearly 70-year presence of Palestinian refugees in Arab host countries. In Lebanon, Palestinians are often accused of instigating the Lebanese civil war (Hudson, 1978). First authorized to live in tented settlements, they took advantage of the presence of the Palestine Liberation Organization in the early 1970s, and gradually adapted their settlements into permanent and armed ghettos outside the control of the Lebanese state (Doraï, 2006). Thus, by adopting a no-camp policy, the Lebanese Government was seeking to avoid the experience of the Palestinian camps, which have evolved over the years into extraterritorial sites of radicalization and armed resistance (Hanafi & Long, 2010).

3.2 The settlement policy: Camps or no camps

Since the beginning of the crisis, political factions in Lebanon have espoused different views on the establishment of organized refugee camps (Thibos, 2014). The most influential political party, the Hezbollah, allied with the Syrian Alawite regime, has openly expressed its concern that formal camps could provide safe havens for Syrian rebels. Other political parties, especially Christian ones, have voiced concern that temporary refugee populations could become permanent residents, altering the precarious sectarian balance in the country (Turner, 2015). These views were not shared by the Sunnis, who have the same religious tradition as most refugees and are generally opposed to the Syrian regime. Finally, most Lebanese political parties have adopted a no-camp policy, fearing the militarization of camps and the radicalization of Syrian refugees, as well as the potential for refugee populations to become permanent (Turner, 2015).

From an economic perspective, Syrians make up the bulk of the unskilled (and often informal) Lebanese workforce (Chalcraft, 2007). Before the Syrian conflict, there were already almost 400,000 Syrian workers in Lebanon, according to estimates (World Bank, 2013). Following the Lebanese civil war, Lebanon and Syria signed the Agreement for Cooperation and Economic and Social Coordination in 1993, benefitting the Syrian regime, then in a position of power. The Agreement called for "freedom of movement of persons between the two countries" and "freedom to reside, work, employ and engage in economic activity in accordance with the laws and regulations in force in each country" (Syrian–Lebanese Higher Council, 1993). It further stated that migrant workers in both countries should receive the same legal treatment as local workers. Although large numbers of Syrians left Lebanon following the assassination of Prime Minister Rafiq Al-Hariri in 2005 and the Israeli bombardments in 2006, Syrians still constituted between 20% and 40% of Lebanon's unskilled labor force in the early 2000s, according to estimates (Chalcraft, 2009).

Thus, the Government's refusal to build organized camps for Syrian refugees has both security and economic rationales. It was possibly designed to serve the economic need for large numbers of low-wage workers by expanding the labor supply, while lowering wages and eschewing worker protections (Turner, 2015). Lebanese capital owners—employers, large agriculture producers, traders, landlords, and construction contractors—are the clear economic beneficiaries of the non-encampment policy (International Labour Organization, 2014). Yet, apart from this group, the Lebanese society has been negatively impacted by the presence of large numbers of vulnerable Syrian refugees. Its economy, housing sector, infrastructure, and social services are all under severe strain.

3.3 Impacts of hosting refugees outside camps

Before the Syrian crisis, Lebanon had one of the highest rates of income inequality in the world, with an especially high poverty rate (Assouad, 2018). Settling large numbers of refugees in poor communities has imposed enormous challenges on the country and particularly on vulnerable host communities (International Labour Organization, 2014). Between 2011 and 2014, the supply on the job market was 30% higher, thus decreasing wages and doubling the unemployment labor rate in an unregulated market (International Labour Organization, 2014; World Bank, 2013; Zetter & Ruaudel, 2014). Annual GDP growth fell from an average of 9% between 2007 and 2010 to less than 2% in 2013. Private investment was curtailed and cross-border trade, real estate, and tourism—important sectors in the Lebanese economy—all suffered (World Bank, 2013). As of 2017, nearly 30% of the Lebanese population and 70% of the Syrian refugees live below the poverty line (Government of Lebanon & United Nations, 2019).

Competition over poorly paid jobs is one of the most serious problems in the country. Syrian workers are willing to work longer hours and with fewer benefits than Lebanese ones (Centre Libanais des Droits Humains, 2013; International Labour Organization, 2014). Moreover, due to purchasing-power-parity disparities between the two countries, Syrians accept wages that are one-third lower than typical Lebanese wages (UNICEF, OCHA, & REACH, 2015), inducing a loss of income for Lebanese and Syrians alike (Loveless, 2013). Humanitarian assistance is helping to stabilize the situation, but aid is not sufficient to improve livelihoods, and Syrians often find themselves forced to work in difficult conditions (Fawaz et al., 2014). Consequently, most refugees work illegally, which increases their vulnerability and makes them an easier target for exploitation (Turner, 2015; Zetter & Ruaudel, 2014).

The huge inflow of refugees in host communities is also placing an additional burden on healthcare and education services, and both sectors are struggling to cope with excess demand (Dahi, 2014; International Labour Organization, 2014; Loveless, 2013). In Lebanon, the Government does not provide most social services. Instead, informal networks, such as families and religious institutions, provide for their fellow citizens (Ward, 2014). Furthermore, large refugee populations have strained the water and power supply in the country, which has struggled to meet the demand for decades. Indeed, many Lebanese rely on private resources to get electricity and water due to the Government's ineptitude to provide basic utilities (Zetter & Ruaudel, 2014).

Initially, the Lebanese population was very welcoming and generous to Syrian refugees. Almost 15,000 people were accommodated free of charge in the homes of Lebanese families and friends. But as the situation became more protracted and resources dwindled, other solutions had to be found. Different types of settlements emerged, informally scattered within communities in almost 2000 locations (Hidalgo et al., 2015; Kikano, Fayazi, & Lizarralde, 2015). More than 60% of the refugees are in the Bekaa, North, and Mount Lebanon regions, where some villages have seen their numbers double or triple (Dahi, 2014; Lenner & Schmelter, 2016).

More than 80% of refugees have resorted to informal rented accommodations, where they are compelled to pay high prices for small dwellings, often shared with other families (Dahi, 2014; UNICEF et al., 2015). Affordable housing is rare and many refugees have been forced to move to less costly accommodations with unsuitable conditions (Government of Lebanon &

United Nations, 2019; International Labour Organization, 2014; Loveless, 2013). Almost 60% of the refugees live in rented apartments or rooms, 18% live in the almost 1500 informal tented settlements, and the rest live in unfinished buildings, nonresidential structures, Palestinian camps, collective shelters, and a few organized camps built by a group of NGOs on private land (Kikano, 2015). Moreover, 24% live in substandard buildings and 12% of the shelters are considered to be in dangerous conditions (Government of Lebanon & United Nations, 2019; Kikano & Lizarralde, 2018; UNHCR, UNICEF, & World Food Programme, 2016).

The large number of refugees has provoked a dramatic surge in demand that no regular housing market could absorb (Hidalgo et al., 2015). But Lebanese owners are mostly responsible for rent inflation. They often replace Lebanese tenants with Syrians who pay higher rents due to the fact that Syrian families are willing to share a single accommodation. In many cases, apartments are illegally divided into rooms with makeshift kitchens and toilets. These spaces are usually precarious, unsound, and overcrowded. The agreements between owners and refugees are often informal. With few resources, refugees often struggle to pay rent and suffer from frequent evictions, as is the case of Syrian refugees living in Nabaa'. Conversely, Lebanese residents in such areas feel their territory has been expropriated by refugees, a situation that contributes to intercommunal tensions.

In rural areas, some settlements, used to house seasonal agriculture workers from Syria, predate the crisis. Tents erected to host workers' family members and friends proliferated very quickly. Parkinson (2014) argues that these settlements are now remaking the "human geography" of the country. Refugees in this type of settlement suffer from climatic inadaptability, lack of privacy, overcrowding, and frequent evictions, as claimed by Syrian refugees in Kab Elias. Moreover, these settlements are a source of perilous environmental problems: toilet units provided by UNHCR are not connected to the public sewer system and the wastewater pollutes agriculture land and the water table (Kikano et al., 2015). Agriculture production is also stymied as Syrian refugees are renting agricultural lands to build their tents, a solution favored by Lebanese landowners, who generate greater income from this practice (Fawaz et al., 2014).

To deal with the housing problem, some political leaders still believe that hosting refugees in camps is a viable solution, as explained by an interviewed UNHCR representative. But while the Lebanese Government recently favored the establishment of camps in Syria or close to the border—a new position in response to the protracted crisis—, UNHCR is calling for the establishment of camps in safe areas within Lebanon. However, studies show that it is too late to erect camps in Lebanon to house refugees for political, social, and economic reasons (Fawaz et al., 2014).

NGOs and private organizations are instead proposing alternative solutions to improve refugees' living environments. These solutions include the following: the IKEA model recommended by UNHCR; Ghata, an adaptable and transportable temporary shelter designed by the American University of Beirut; the International Humanitarian Relief's shelter in Arsal; and the Shelter Box Plan designed by the Danish Refugee Council. The Lebanese Government has rejected most of these for two main reasons. First, because temporary shelters are seen as a "threat to the nation" due to their potential permanence (Yassin, Osseiran, Rassi, & Boustani, 2015); and second, because they would create feelings of resentment among Lebanese communities who reside in poor housing conditions.

As the Government has refused to build camps or consider alternative solutions, NGOs can only improve existing shelters. The improvements include consolidating the structure and replacing materials—long-lasting materials are prohibited—, which require ongoing maintenance (Fawaz et al., 2014). Meanwhile, despite governmental restrictions, temporary solutions are becoming permanent and urban and rural areas are undergoing informal and often durable transformations, as observed in my fieldwork.

Nevertheless, most Syrian refugees continue to live in very precarious circumstances. Refugees are now generally perceived as the cause of all social and economic ills—40% of the Lebanese population report negative to very negative feelings toward them (UNICEF et al., 2015). Deprived of protection, Syrian refugees often face arbitrary arrests, experience violence and racism, and have their fundamental rights violated—prejudices that the Government has done little to stop (Centre Libanais des Droits Humains, 2013). Tensions between both communities are also rising due to the sectarian nature of the Syrian conflict and its impact on denominational and political divisions within Lebanon. The Lebanese population generally fears that the influx of Syrian refugees constitutes another form of occupation, similar to the invasion by the Syrian army for almost 20 years (Naufal, 2012; UNICEF et al., 2015).

3.4 The 2014 regulations

Excluding refugees from formal governance systems does not compel them to return to their home country. Yet, most host nations, especially those with fragile governance systems like Lebanon, keep adopting this strategy (Kibreab, 2003).

In 2013, in an attempt to reduce competition over jobs, the Lebanese Government began restricting work permits for Syrian refugees. Despite a Syrian population of almost 1.5 million, the Government only issued or renewed 1233 work permits for Syrian citizens that year (International Labour Organization, 2014). Due to funding shortages, only the most vulnerable refugees—44% out of 70% of the Syrians registered with UNHCR—receive financial aid, and many refugees need to work to survive. Since less than 1% hold work permits, most Syrians work illegally and are exposed to exploitation and abuse (Government of Lebanon & United Nations, 2019).

In October 2014, Lebanon's Council of Ministers adopted the first comprehensive policy addressing Syrians' displacement. The new regulations sought to dissuade Syrian refugees and Palestinian refugees from Syria from seeking protection in Lebanon, with the goal of decreasing their number and protecting the country's economy and social stability (Onishi, 2013; Turner, 2015). Firstly, the Government prohibited UNHCR from registering new refugees. Secondly, it replaced the open-border policy with new visa and residency regulations, which went into effect in January 2015. These regulations define two categories of Syrian refugees: those registered with UNHCR and those with Lebanese sponsors who guarantee their livelihoods. The former lose their right to work, whereas the latter can only work in certain fields. For both categories, the biannual renewal of the residence permits (for persons over 15 years old) costs USD200. The regulations also tightened rules for newcomers. As most refugees are unable to renew their permit, almost 70% of the Syrian refugees do not have legal status in Lebanon (Janmyr, 2016; UNHCR et al., 2016).

These regulations also stipulate that Syrian refugees with no residency permit may be detained and forced to return to Syria. Fear of arrest amidst ongoing social and political tensions has forced most of them to go underground (Norwegian Refugee Council, 2014; Onishi, 2013). Lack of legal safeguards also limits refugees' work opportunities. Most lost their access to social services and healthcare and the moderate protection they once enjoyed (Lenner &

Schmelter, 2016). Although the Ministry of Education does not require residency permits for school children, fewer Syrian children are now attending schools in Lebanon (UNHCR et al., 2016).

According to Van Hear (2006), this policy of exclusion deprives refugees of their basic rights and disempowers them, limiting their ability to relocate or to return home. In fact, the institutional exclusion of Syrian refugees has not reduced their numbers in Lebanon. Instead, it has increased the likelihood of workplace exploitation (lower pays, longer hours, and more hazardous conditions), while reducing the possibility of legal recourse, as evidenced during group discussions with Syrian refugees in several localities.

4. Policy recommendations

In the past decade, due to declining humanitarian budgets, many organizations have combined humanitarian assistance with local development programs to manage refugee influxes and support poor communities hosting them. UNHCR adopted a humanitarian–development approach in 2009 (UNHCR, 2009), based on the fact that almost 70% of the world's refugees live outside organized camps and more than 65% of the refugee situations become protracted (Ward, 2014). Yet, studies have shown that when refugees are empowered through their economic integration into host communities, both groups are more likely to thrive (Betts, Bloom, Kaplan, & Omata, 2014; Collier & Betts, 2017).

My analysis of other protracted situations in developing countries (Betts et al., 2014; Jacobsen & Fratzke, 2016; Omata, 2017), particularly the Jordanian Government's response to the Syrian refugee crisis (Collier & Betts, 2017), inspired the policy recommendations in this chapter. In 2016, the Jordanian Government adopted an inclusive policy, facilitating the procurement of work permits and formalizing existing jobs for Syrian refugees. The policy was restricted to certain fields and applied according to specific conditions, so as not to create competition between Syrians and Jordanians in the labor market, yet still giving Syrian refugees a chance to thrive. In exchange, Jordan received international funds to invest in development projects.

My recommendations are based on two complementary ideas. First, Lebanon *should* benefit from hosting large numbers of refugees. The Lebanese Government, with the support of its international and national partners, should take control of the refugee situation and become the main recipient and manager of humanitarian funds. These funds could be channeled into large-scale development projects to upgrade dilapidated infrastructure and social services. Second, the Syrian refugees' situation in terms of legal status, jobs, and housing needs to be temporarily regularized. The proposed solutions are detailed below; however, these core concepts still require insights from economists, political scientists, urban planners, and other related experts.

4.1 The development strategy led by the Lebanese Government

In most refugee situations, donors allocate funds to host governments, out of respect for their sovereignty (Agnew, 2005). Yet in Lebanon, donors have directly channeled funds to humanitarian organizations. These organizations have not involved the Government in their efforts for many reasons, including protracted periods of political uncertainty, increased regional instability, corruption, and poor fiscal governance. Hence, UNCHR and its partners have until now received, managed, and controlled most funds for the Lebanese refugee crisis.

Development initiatives exist, but they are localized and cannot address the issues at scale. In fact, the absence of a national policy has produced a decentralized response and the emergence of many municipalities as authoritative bodies (Fawaz et al., 2014; Government of Lebanon & United Nations, 2019). Although local authorities have been effective at addressing local needs, there are a number of drawbacks with this approach. Areas with more welcoming policies have experienced a huge influx in refugees and, consequently, are facing a heavier burden. Many of these municipalities have established agreements with humanitarian agencies and welcomed development initiatives funded by these organizations, but such initiatives cannot address capacity issues, as they remain at a local level.

The scale of these development projects is not the only issue. Large-scale development projects need regular and substantial international funds. Lebanon should receive regular funding according to its needs as it is shouldering the bulk of the Syrian refugees' burden. Like most developing countries, it should strategically "use" this argument to request more funding from potential donors (Harrell-Bond, 1998). Presently, this level of funding is not available. According to UNHCR, Lebanon has received nearly 50% of the aid it needs to address the refugee crisis effectively (Kelley, 2017).

Three main reasons explain the funding shortage. First, the global funding gap, which according to UNHCR is "unable to meet the basic needs of millions" (Grant, 2015). Second, the international community's declining interest in the Syrian refugee crisis, a recurrent pattern in refugee crises (Kelley, 2017). Third, the reluctance of some Lebanese ministers to accept refugee-related funds, fearing that these could lead to the naturalization of more Syrian expatriates (Minority Rights Group International, 2019).

But international funds are of the upmost importance for Lebanon. Successive governments have not been able to implement large-scale projects to rehabilitate dilapidated infrastructure and social services. Instead, the country is among the most indebted in the world (World Bank, 2013). If Lebanon manages this crisis to its advantage, replicating the main principles adopted in Jordan, international assistance could help fund these development projects. Indeed, hosting Syrian refugees in Lebanon could strengthen the state's governance.

To prevent mismanagement and corruption given Lebanon's poor record of fiscal governance, a committee composed of representatives from the Lebanese line ministries, donor countries, municipalities, and the host and refugee communities would manage funds. The stakeholders would be chosen on the basis of their specialization in the national sectors in need of development. The committee's primary role would be to elaborate on an action plan for development initiatives. It would also specify employment quotas for refugees. To limit competition with the local population, Syrians would be encouraged to work in projects and sectors usually dominated by foreign workers, whose numbers are declining due to the ongoing fiscal crisis (El-Hage, 2019). Increasing the number of jobs available to refugees, while decreasing competition between Syrian refugees and Lebanese workers, would ease tensions between both communities.

4.2 Temporary regularization of the Syrian refugees' situation

As mentioned earlier, Syrian refugees are exposed to exploitation due to their precipitous legal status; they often work in the informal job market in substandard conditions and compete over unskilled jobs with poor Lebanese. Therefore, a rights-based transparent legal status for Syrians would not only grant them more protection but also make their stay less detrimental to the country's economy, security, and social stability.

Moreover, affordable residency and work permits would constitute a significant source of income for the Government. Another source of income would come from taxes imposed on Syrian migrants or refugees who temporarily, but legally, reside and/or work in Lebanon. In addition, if employment is organized instead of restricted, and if large-scale projects benefitting infrastructure improvement, agricultural sectors, and environmental works are created, Lebanese citizens and refugees alike would benefit from new job opportunities. After legalizing jobs for refugees, the Government could institute new wage policies to reduce income inequality, decrease housing prices, increase purchasing power, and reduce the unemployment rate. Refugees could then be seen as a productive force with the potential to benefit the Lebanese economy, rather than harm it.

However, unemployed refugees are generally perceived to be a burden and employed refugees are seen as competing with host communities for jobs (Chambers, 1986; Zetter, 1992). To ease tensions between communities, the Government should encourage refugees' participation in sectors of the job market that do not face competition from Lebanese citizens. Given the large number of foreign workers in Lebanon, this could be easily implemented. Meanwhile, refugees holding a professional degree should be allowed to practice their profession according to certain regulations (passing academic tests, accrediting diplomas, etc.). Syrians in the formal labor market would no longer be eligible to receive financial aid, alleviating pressure on humanitarian-aid agencies. The number of aid workers would be significantly reduced and humanitarian funds could be invested instead in development initiatives.

As for housing, rent arrangements between landlords and Syrian tenants should be formalized and rent deals should be registered in municipalities. This measure would protect the rights of refugees and Lebanese tenants by mitigating rent increases. Formalizing refugee-housing deals would also prevent refugees from transforming existing structures into substandard dwellings. These transformations are turning poor but organized urban neighborhoods into chaotic shantytowns. Formal agreements would also facilitate greater use of agricultural land. Settlements could be built on nonagriculture lots, with core houses connected to public infrastructure, allowing refugees to develop them according to their needs and aspirations, with reference to predesigned architectural models and prechosen materials.

To conclude, I recommend the following: (1) The centralization of the Lebanese state's role in managing the refugee crisis—without excluding municipalities; (2) Donors' allocation of adequate and regular funding to Lebanon for the refugee crisis; (3) Allocation of donor funds directly to the Lebanese Government, for the management of development projects to be overseen by a multi-stakeholder committee; (5) Adoption of a development strategy by the Lebanese Government aimed at upgrading dilapidated infrastructure and social services; and (6) Temporary inclusion of Syrian refugees in formal systems (Fig. 11.1).

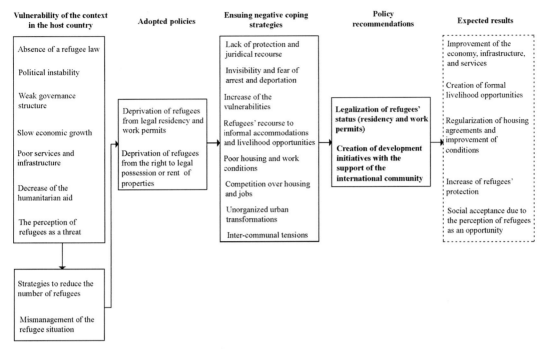

FIGURE 11.1   From informality to the regularization of Syrian refugees' situation. *Credit: Faten Kikano (2020).*

<hr/>

5. Conclusion

<hr/>

Roth (2004) claimed that any refugee crisis is one of politics, not one of capacity. Although what he stated is generally valid, it is not applicable to Lebanon, which is presently witnessing one of the worst financial and political setbacks since its independence in 1943. The Lebanese situation is being aggravated by the nine-year-long refugee crisis. But while analyses show that the Syrian conflict will eventually last and that the number of Syrians in Lebanon is likely to remain stable (Government of Lebanon & United Nations, 2019), the Lebanese authorities still lack the will to perceive this crisis as anything but temporary.

This study highlights two main problems. First, the Lebanese state is excluding refugees from formal systems with the intention of pressuring them out of its territory. Second, countries in the Global North have been making every effort to keep Syrians away from their borders. Only 5% of the Syrian refugees are hosted outside of Syria's neighboring countries (Amnesty International, 2020). Various international and national political parties have expressed their concern at the amplified risks of dragging Lebanon into the Syrian conflict due to hosting a large refugee population (Naufal, 2012). Yet, despite the fact that the Lebanese state, not having signed the 1951 Refugee Convention, has no legal responsibility in hosting refugees, Western countries are still putting a lot of pressure for containing refugees in Lebanon—and in other neighboring countries—to mitigate hosting them in their homeland (Ward, 2014). Simultaneously, the donor countries are imposing significant cutbacks on humanitarian funds and only a limited number of refugees, identified as the most vulnerable ones, receive aid.

Syrian refugees often integrate informal systems to ensure their basic needs in relation to housing and livelihood. In the absence of regularization of their legal status, housing solutions, and jobs, they become more vulnerable and are unable to leave the host state and to relocate to their home country or another country. Weakened on many levels, they accept poor work and housing conditions, replacing in many sectors the local communities. As a result, both communities become poorer and social tensions between them grow. Refugees, with limited protection and also limited control from the state, become invisible settlers. Paradoxically, while their invisibility exposes them to many hazards, their presence may constitute a security risk for the country.

In response to these issues, this study sets out a number of policy recommendations, which focus on two intertwined points. First, the replacement of the humanitarian approach with a development strategy that would allow the upgrade of infrastructure and service systems in Lebanon, leading to the creation of jobs for refugees and their vulnerable Lebanese hosts. In addition to its positive outcomes for the country, the proposed approach would alleviate the ongoing need of decreasing and insufficient humanitarian aid. It would also lead to a change of paradigm in the perception of refugees by the host state and the host communities: Instead of being seen as a burden, refugees would be perceived as populations that motivate large investments benefitting the country's systems and economy.

The second recommendation point is the regularization of the Syrians refugees' situation and their temporary integration in the legal systems. Their integration would lead to their protection as well as to that of the communities hosting them. Their empowerment would be the best incentive for their return to Syria or for their resettlement in another country, capable of offering them decent, stable, and secure living conditions. The formalization of their status would also be a source of revenue for the Lebanese Government, through payments of residency documents, work permits, and taxes on their income.

The suggested policy recommendations—especially regarding the temporary integration of refugees in formal systems—would likely pose a number of challenges due to cultural differences between Lebanese and Syrians, and also due to the fear of certain religious communities of the loss of their identity following significant demographic transformations. Yet, reassessing the Syrian refugees' situation and implementing adapted solutions have become vital for both communities and the socioeconomic stability of the country as a whole.

Acknowledgments

The author would like to thank Hannah Kikano for her support for the editing of this chapter.

References

Agnew, J. (2005). Sovereignty regimes: Territoriality and state authority in contemporary world politics. *Annals of the Association of American Geographers, 95*(2), 437–461.

Amnesty International. (2020). *Syria's refugee crisis in numbers*. Retrieved from https://www.amnestyusa.org/syrias-refugee-crisis-in-numbers.

Assouad, L. (2018). *Rethinking the Lebanese economic miracle: The extreme concentration of income and wealth in Lebanon.* World Inequality Database working paper 2017/13. Retrieved from https://wid.world/document/rethinking-lebanese-economic-miracle-extreme-concentration-income-wealth-lebanon-2005-2014-wid-world-working-paper-201713.

Betts, A., Bloom, L., Kaplan, J. D., & Omata, N. (2014). *Refugee economies: Rethinking popular assumptions*. Oxford: University of Oxford.

Centre Libanais des Droits Humains. (2013). *Travailleurs syriens au Liban. Une évaluation de leurs droits et de leur réalité* [*Syrian workers in Lebanon: An evaluation of their rights and reality*]. Retrieved from http://www.rightsobserver.org/files/Report_Syrian_Workers__FR_final.pdf (in French).

Chalcraft, J. (2007). Labour in the levant. *New Left Review, 45*, 27−47.

Chalcraft, J. (2009). *The invisible cage: Syrian migrant workers in Lebanon*. Stanford, CA: Stanford University Press.

Chambers, R. (1986). Hidden losers? The impact of rural refugees and refugee programs on poorer hosts. *International Migration Review, 20*(2), 245−263.

Collier, P., & Betts, A. (2017). *Refuge: Rethinking refugee policy in a changing world*. New York, NY: Oxford University Press.

Crépeau, F., Nakache, D., & Atak, I. (2007). International migration: Security concerns and human rights standards. *Transcultural Psychiatry, 44*(3), 311−337.

Dahi, O. (2014). The refugee crisis in Lebanon and Jordan: The need for economic development spending. *Forced Migration Review, 47*, 11−13.

Doraï, M. K. (2006). *Les réfugiés palestiniens du Liban. Une géographie de l'exil* [*The Palestinian refugees in Lebanon: A geography of exile*]. Paris: CNRS (in French).

El Mufti, K. (2014). *Official response to the Syrian refugee crisis in Lebanon, the disastrous policy of no-policy*. Report of the Civil Society Knowledge Centre. https://doi.org/10.28943/CSKC.002.20001.

El-Hage, A. (December 5, 2019). Migrant workers hit hard by the dollar crisis in Lebanon. *L'Orient le Jour*. Retrieved from https://www.lorientlejour.com/article/1197481/migrant-workers-hit-hard-by-the-dollar-crisis-in-lebanon.html.

Fawaz, M., Saghiyeh, N., & Nammour, K. (2014). *Housing, land and property issues in Lebanon: Implications of the Syrian refugee crisis*. Beirut: UNHCR & UN-Habitat.

Government of Lebanon, & United Nations. (2019). *Lebanon crisis response plan (LCRP) 2017−2020*. Retrieved from https://www.unhcr.org/lb/wp-content/uploads/sites/16/2019/04/LCRP-EN-2019.pdf.

Grant, H. (September 6, 2015). UN agencies 'broke and failing' in face of ever-growing refugee crisis. *The Guardian*. Retrieved from https://www.theguardian.com/world/2015/sep/06/refugee-crisis-un-agencies-broke-failing.

Hanafi, S., & Long, T. (2010). Governance, governmentalities, and the state of exception in the Palestinian refugee camps of Lebanon. *Journal of Refugee Studies, 23*(2), 134−159. https://doi.org/10.1093/jrs/feq014.

Harrell-Bond, B. (1998). Camps: Literature review. *Forced Migration Review, 2*, 22−23.

Hidalgo, S., LaGuardia, D., Trudi, G., Sole, R., Moussa, Z., Van Dijk, J., ... Zimmer, L. (2015). *Beyond humanitarian assistance? UNHCR and the response to Syrian refugees in Jordan and Lebanon, January 2013 − April 2014*. Evaluation report. Retrieved from https://www.refworld.org/docid/559f9b1a4.html.

Hudson, M. C. (1978). The Palestinian factor in the Lebanese civil war. *The Middle East Journal, 32*(3), 261−278.

International Labour Organization. (2014). *Assessment of the impact of Syrian refugees in Lebanon and their employment profile*. Retrieved from https://www.ilo.org/beirut/publications/WCMS_240134/lang−en/index.htm.

Jacobsen, K., & Fratzke, S. (2016). *Building livelihood opportunities for refugee populations: Lessons from past practice*. Migration Policy Institute report. Retrieved from http://www.migrationpolicy.org/research/building-livelihood-opportunities-refugee-populationslessons-past-practice.

Janmyr, M. (2016). Precarity in exile: The legal status of Syrian refugees in Lebanon. *Refugee Survey Quarterly, 35*(4), 58−78. https://doi.org/10.1093/rsq/hdw016.

Kelley, N. (2017). Responding to a refugee influx: Lessons from Lebanon. *Journal on Migration and Human Security, 5*(1), 82−104.

Kibreab, G. (2003). Displacement, host governments' policies, and constraints on the construction of sustainable livelihoods. *International Social Science Journal, 55*(175), 57−67.

Kikano, F. (September 6, 2015). *Camps or no camps: Housing and sheltering solutions for/by Syrian refugees in Lebanon*. Paper presented at the Refuge from Syria Conference, Refugee Study Center, Oxford, United Kingdom.

Kikano, F., Fayazi, M., & Lizarralde, G. (2015, July). *Understanding forms of sheltering by (and for) Syrian refugees in Lebanon*. Paper presented at the 7th i-Rec Conference 2015: Reconstruction and Recovery in Urban Contexts, London. Retrieved from http://www.grif.umontreal.ca/i-Rec2015/RT2A_18_Kikano%20et%20al_Understanding%20Forms%20of%20Sheltering.pdf.

Kikano, F., & Lizarralde, G. (2018). Settlement policies for Syrian refugees in Lebanon and Jordan: An analysis of the benefits and drawbacks of organized camps. In A. Asgari (Ed.), *Resettlement challenges of refugees and disaster displaced populations* (pp. 29−40). Cham, Switzerland: Springer.

Lenner, K., & Schmelter, S. (2016). Syrian refugees in Jordan and Lebanon: Between refuge and ongoing deprivation. In IEMed (Ed.), *IEMed Mediterranean Yearbook 2016* (pp. 122–126). Barcelona: IEMed.

Loveless, J. (2013). Crisis in Lebanon: Camps for Syrian refugees? *Forced Migration Review, 43*, 66–68.

Minority Rights Group International. (2019). *An uncertain future for Syrian refugees in Lebanon: The challenges of life in exile and the barriers to return.* Retrieved from https://minorityrights.org/publications/an-uncertain-future-for-syrian-refugees-in-lebanon-the-challenges-of-life-in-exile-and-the-barriers-to-return.

Naufal, H. (2012). *Syrian refugees in Lebanon: The humanitarian approach under political divisions.* Migration Policy Center Research Report 2012/13. Retrieved from http://cadmus.eui.eu//handle/1814/24835.

Norwegian Refugee Council. (2014). *The consequences of limited legal status for Syrian refugees in Lebanon.* Retrieved from https://www.nrc.no/resources/reports/the-consequences-of-limited-legal-status-for-syrian-refugees-in-lebanon.

Omata, N. (2017). *Who thrives, who struggles? Exploring the determinants of economic success among refugees.* Retrieved from https://www.qeh.ox.ac.uk/blog/who-thrives-who-struggles-exploring-determinants-economic-success-among-refugees.

Onishi, N. (December 11, 2013). Lebanon worries that housing will make Syrian refugees stay. *New York Times.* Retrieved from http://www.nytimes.com.

Parkinson, S. E. (April 3, 2014). *Refugee 101: Palestinians in Lebanon show refugees from Syria the ropes.* Retrieved from https://merip.org/2014/04/refugee-101.

Roth, K. (2004). Defending economic, social and cultural rights: Practical issues faced by an international human rights organization. *Human Rights Quarterly, 26*(1), 63–73.

Syrian–Lebanese Higher Council. (1993). *Agreement for Economic and Social Cooperation between the Lebanese Republic and the Syrian Arab Republic.* Retrieved from http://www.syrleb.org/SD08/msf/1507751474_.pdf.

Thibos, C. (2014). *One million Syrians in Lebanon: A milestone quickly passed. Migration policy center policy brief 2014/03.* Retrieved from http://cadmus.eui.eu/bitstream/handle/1814/31696/MPC_THIBOS_2014_.pdf.

Turner, L. (2015). Explaining the (non-)encampment of Syrian refugees: Security, class and the labour market in Lebanon and Jordan. *Mediterranean Politics, 20*(3), 386–404.

UNHCR. (2009). *UNHCR policy on refugee protection and solutions in urban areas.* Retrieved from https://www.unhcr.org/protection/hcdialogue%20/4ab356ab6/unhcr-policy-refugee-protection-solutions-urban-areas.html.

UNHCR. (2011). *UNHCR global appeal 2010–2011 – Lebanon.* Retrieved from https://www.unhcr.org/publications/fundraising/4b05121f9/unhcr-global-appeal-2010-2011-lebanon.html.

UNHCR. (2016). *3RP, Regional and Refugee Resilience Plan 2015–2016 in response to the Syria crisis. Report.* Retrieved from http://reporting.unhcr.org/node/12589.

UNHCR, UNICEF, & World Food Programme. (2016). *Vulnerability assessment of Syrian refugees in Lebanon 2016.* Retrieved from https://reliefweb.int/sites/reliefweb.int/files/resources/VASyR2016.pdf.

UNICEF, OCHA, & REACH. (2015). *Defining community vulnerabilities in Lebanon.* Retrieved from https://data2.unhcr.org/en/documents/download/44875.

Van Hear, N. (2006). 'I went as far as my money would take me:' Conflict, forced migration and class. In F. Crépeau, D. Nakache, M. Collyer, H. N. Goetz, & A. Hansen (Eds.), *Forced migration and global processes: A view from forced migration studies* (pp. 125–158). Oxford: Lexington Books.

Ward, P. (2014). Refugee cities: Reflections on the development and impact of UNHCR urban refugee policy in the Middle East. *Refugee Survey Quarterly, 33*(1), 77–93. https://doi.org/10.1093/rsq/hdt024.

World Bank. (2013). *Lebanon: Economic and social impact assessment of the Syrian conflict.* Retrieved from http://documents.worldbank.org/curated/en/925271468089385165/Lebanon-Economic-and-social-impact-assessment-of-the-Syrian-conflict.

Yassin, N., Osseiran, T., Rassi, R., & Boustani, M. (2015). *No place to stay? Reflections on the Syrian refugee shelter policy in Lebanon.* UN-Habitat and IFI/AUB report. Retrieved from https://www.aub.edu.lb/ifi/Documents/publications/books/2015-2016/20150907_noplacetostay.pdf.

Zetter, R. (1992). Refugees and forced migrants as development resources: The Greek Cypriot refugees from 1974. *Cyprus Review, 4*(1), 7.

Zetter, R., & Ruaudel, H. (2014). Development and protection challenges of the Syrian refugee crisis. *Forced Migration Review, 47*, 6–10.

12

Incremental housing in Villa Verde, Chile: A view through the Sendai Framework lens

David O'Brien, Sandra Carrasco

Faculty of Architecture, Building, and Planning, University of Melbourne, Melbourne, Victoria, Australia

Enhancing Disaster Preparedness
https://doi.org/10.1016/B978-0-12-819078-4.00012-5

223

1. Introduction

Researchers, community leaders, and general observers often claim that resilient communities cope with "natural" and human-made disasters in ways that significantly improve their recovery processes (Carrasco, Ochiai, & Okazaki, 2016; Davis, 2013; Lizarralde, 2009; Oliver-Smith, 1991; O'Brien & Ahmed, 2012; Schilderman, 2004). Thus, by empowering resilient communities and leveraging their capacities to rebound from disasters, governments and other support networks can promote households' quick and efficient recovery. Likewise, private-sector enterprises have a stake in resilient recovery processes as they also suffer from losses to productivity and facilities, and with impacts on their workers' housing and livelihoods.

The Villa Verde settlement in Chile exemplifies this type of inclusive approach. Following the 8.8-magnitude earthquake and tsunami that devastated many coastal cities in Chile in 2010, the city of Constitucion, where 500 people died, lost nearly two-thirds of its housing stock (Fig. 12.1). The Chilean housing policy previously followed a top-down approach. However, the Villa Verde settlement embraced lessons from informal-housing procurement models, which encourage residents to take an active role in developing their houses. Aside from easing the financial burden on the government, this strategy reduces the economic risks faced by residents, who take control of their housing improvements according to their financial capacities. They are also able to develop their houses based on personal needs and aspirations.

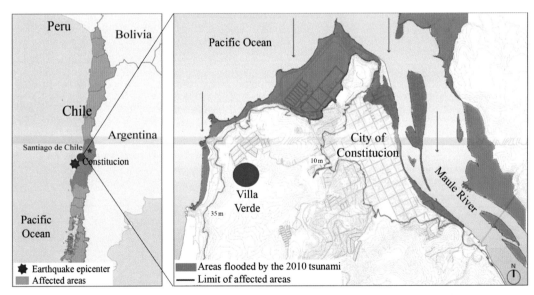

FIGURE 12.1   Location of Constitucion and Villa Verde, in Chile. *Credit: The authors, based on MINVU (Ministerio de Vivienda y Urbanismo), Municipio de Constitución, & Arauco. (2010).* Plan de Reconstrucción Sustentable PRES Constitución [Plan for the sustainable reconstruction of Constitucion]. *Retrieved from http://minvuhistorico.minvu.cl/ incjs/download.aspx?glb_cod_nodo=20100910140027&hdd_nom_archivo=Plan   Maestro_PRES   ConstituciónB3n_Agosto 2010.pdf (in Spanish).*

This chapter highlights a process, defined as incremental housing, whereby households in the Villa Verde settlement took a proactive role in rebuilding their lives and homes during a recovery phase that is open-ended and informal. Supported by policies that encouraged urbanization away from tsunami-prone areas and the use of earthquake-resistant lightweight materials, and subsidized by industry and government, Villa Verde was praised for its comprehensive approach to development. However, an in-depth study of the project as it evolved revealed that the incremental-development process did not necessarily reduce the overall risks faced by the households, by exposing vulnerable communities to another risk type.

In the next section, we discuss the emergence of incremental-housing advocacy and some of its formal practices, with a particular focus on the Latin American and Chilean contexts. Then we present, in Section 3, the methodology adopted in this research. Section 4 concentrates on the Villa Verde case, presenting the project, analyzing the process, and discussing its outputs, including some drawbacks related to disaster risk reduction (DRR). In Section 5, we examine the Villa Verde project, taking the Sendai Framework as an analytical lens. In the conclusions, we highlight some lessons learned through this incremental-housing process.

2. The "half-house" concept

The incremental and self-help housing model has been explored since the 1960s. John Turner (1968) was among the pioneers to advocate the development of squatter settlements. He claimed that these initiatives expressed people's freedom of choice as well as their capacity to manage resources and shape their environment. In a later publication, *Freedom to build*, Turner and Fichter (1972) challenged the housing practices prevalent worldwide in the 1950s and 1960s, as well as governmental strategies to bulldoze slums and relocate their residents to industrialized housing projects (Burgess, 1992). Turner's contribution stressed the importance of homeowners' control over the process of housing production. It also highlighted the organic nature of housing that evolved to meet the residents' needs and aspirations. Thus, for Turner and Fichter (1972), housing should be more a verb than a noun, as it impacts people's social and economic wellbeing. Concurrently, John Habraken (1972) proposed alternatives for mass-produced public housing that harnessed the residents' capacities. Habraken's idea was to provide a support structure that would be infilled by the residents during a process whereby they completed their dwellings (Habraken, 1972; Van Hoogstraten, 2000).

Turner's ideas influenced housing policies in Latin America, as it was the first region in the Global South to experience rapid urbanization and it was the birthplace of approaches aiming to legitimize self-help housing strategies common among the urban poor (UN-Habitat, 2011). One of the global milestones of the self-help theories is the experimental housing project known as PREVI in Lima, which was initiated by the Peruvian Government in the 1960s. PREVI was a mass-produced housing project initially based upon prefabricated construction systems and designed to promote resident-driven incremental improvements. The incremental processes of the PREVI houses were documented by García-Huidobro, Torres Torriti, and Tugas (2008), who revealed that the houses accommodated changing social demands and income-generation activities via subletting and commercial use. Turner acknowledged the value of the PREVI initiative as promoting a "community-building activity, not just a product" (García-Huidobro et al., 2008, p. 7). Additionally, in recent decades, researchers have studied incrementally built housing and observed its role in supporting economic sustainability and enhancing place attachment (Friedman, 2002; Khan, 2014; Tipple, 2000).

Architects from a Chilean firm, Elemental, managed by Alejandro Aravena, led the design phase of the Villa Verde settlement. They first drew international attention with their "half-house" model previously implemented at the Quinta Monroy housing estate at Iquique in northern Chile in 2004. The half-house concept was driven by the financial limitation of the governmental subsidy, which was USD10,000 per house. The subsidy was delivered via the Ministry of Housing and Urban Development (MINVU) to private developers or NGOs to construct large-scale settlements (MINVU, 2017). Elemental calculated that such a subsidy was only enough to build "half" of the required house. Thus, Elemental designed small apartments that could double in size, by leaving porous spaces between them to encourage infills by the residents (Aravena, 2010, 2014).

Elemental's half-house concept somehow combines the theories and practices proposed by Turner (1968) and Habraken (1972). Although this strategy was then promoted by Elemental as an innovative solution, the practice of incremental and self-built housing already had an established history in Chile and abroad. For instance, the Chilean capital, Santiago, hosts the Andalucía community developed since 1990 and managed by the MINVU. This community was inserted into the city center in contrast to the standard practice of relocating marginalized and low-income households to the city periphery. The MINVU promoted the construction of 180 houses with a base footprint of 30 m^2, in a volume that could be vertically subdivided into two or three levels and incrementally expanded by up to 70 m^2 through the residents' efforts and at their expenses (Junta de Andalucía, n.d.). The outcomes of this project were highly regarded by the residents and the government alike. The project received multiple awards and was presented in several exhibitions displaying the initiative to both domestic and international architects. After its occupation, the development was improved via an incremental process by the residents who enlarged and consolidated the value of their houses through different types of improvements (Greene, 2017).

Elemental's design tactic was promoted at a time when the Chilean housing policies were under review. This was because many residents abandoned government-managed subsidized housing units and returned to their informal settlements where they were able to manage their houses within self-supporting communities (Morales Martínez et al., 2017; Rodríguez & Sugranyes, 2005; Salvi Del Pero, 2016). Concerns within the MINVU that subsidies were producing such less desirable housing outcomes led the department to reflect on its current practices and to reconsider new strategies that might challenge the status quo.

3. Research design

Elemental has often highlighted the role it played in the redevelopment of the city of Constitucion in the aftermath of the 2010 earthquake and tsunami (Aravena, 2014). Immediately after the disaster, it proposed a water-distribution plan, closely followed by emergency shelters and a masterplan for the city's redevelopment (MINVU, Municipio de Constitución, & Arauco, 2010). Given that Elemental has positioned itself as a key contributor to the post-disaster reconstruction we undertook a thorough analysis of the Villa Verde settlement based on the Sendai

Framework for Disaster Risk Reduction 2015–2030 (UNISDR, 2015). Through the lens of the four Sendai priorities for action, this chapter reviews the outcomes of the project concerning the risks faced by the community, as well as its responses to reduce and accommodate them.

This research was based on a mixed-method approach. The fieldwork took place from August 2017 to October 2019. Data was obtained through archival sources, site surveys, trace analysis, 32 semistructured interviews with residents, and photographic surveys (Groat & Wang, 2013; Yin, 2017; Zeisel, 1984). The authors conducted a comparative review of site plans prepared by Elemental alongside recent site plans. The latter show the extent of the additions implemented by the residents and provide data regarding the scope and location of these works. Based on the analysis of a photographic survey of houses and public spaces, we identified and quantified key types of additions, such as extra rooms, improved fittings, and finishes, as well as other elements—for instance fences, porticos, and carports. The interviews were used to triangulate the collected data, providing us with an additional understanding of how the settlement has evolved and the rationale behind these changes.

4. Incremental housing in Villa Verde

The Villa Verde settlement was designed in 2010 by the Elemental architecture firm and financed by governmental agencies and the timber-processing company Arauco, which holds 80% of the municipality's forested areas (CONAF, 2017). One thousand residents occupied the 484 houses of the first stage as of late 2013 (Fig. 12.2). Arauco's employees received approximately half of the houses. The government reserved the remaining houses for general residents of Constitucion whose dwellings had been destroyed by the 2010 earthquake and tsunami. Some residents were at the same time company employees and eligible for post-disaster housing.

The settlement is located 3 km west of the Constitucion city center on forested land owned by Arauco. It sits adjacent to an established middle-class residential suburb where many of the free-standing houses have been extended to two stories. The mature pine trees on the Villa Verde site were cleared to enable this real-estate operation. The intervention included cutting a series of terraces into the sloping site to provide a level ground for the house clusters. The site planning was dominated by 19 cul-de-sacs (with approximately 400 m^2 each), framed on three sides with rows of attached two-story houses (Fig. 12.3).

Using timber as the main building material, two house typologies were designed: 422 units of Type A and 62 units of Type B, each with a 43-m^2 floor area (Figs. 12.4 and 12.5). During the allocation process, some residents had the opportunity to choose the house type and its location in the settlement. Residents who had attended community meetings at the initial consultation and design stage, predominantly Arauco employees, were able to claim particular houses. On the other hand, residents who came to the project later, mainly disaster-affected people, were allocated the remaining and less desirable Type-B houses. These were typically sandwiched in the middle of longer rows, or provided with reduced access to parking lots, and/or located on sloped sites. As of 2016, Arauco constructed 184 additional houses following the same typologies in the southeastern edge of the site. Residents took possession of the houses of the second stage of the settlement in 2018. In this chapter, our analyses are only focused on the first stage of the Villa Verde project.

FIGURE 12.2   Villa Verde houses before (top) and after (bottom) extensions—the former from the settlement's second stage and the latter from the first stage. *Credit: The authors (2017).*

Based on its experiences in the Quinta Monroy project, Elemental reasoned that rather than designing and building low-quality larger houses in the Villa Verde project, it was preferable to build smaller houses at relatively higher standards. In both projects, the architects' strategy was to design the houses in such a way that encouraged the residents to expand them and complete their other 'halves' over time with their resources, according to their needs and capacities.

In reality, the half-house appellation is somewhat misleading as the houses in Villa Verde were delivered to the residents with a serviced bathroom and kitchen and connected to water, sewerage, and power—all of which expensive elements of the building process. In their delivered state, all houses, each with two bedrooms on the upper story, were ready to be occupied. Type-A houses had a functioning roof over the incomplete half of the house, whereas Type-B houses included a porch area at their side that could be enclosed into the interior space with three new walls. Elemental designed the houses of both types in such a way as to be extended without complex additional work or significant cost for the residents. On the downside, the provided half house was not painted and did not include finished floors, cabinet works, or benchtops.

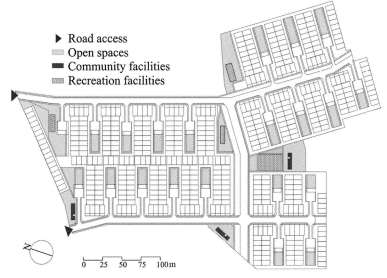

FIGURE 12.3 The layout of the Villa Verde settlement. *Credit: The authors (2020).*

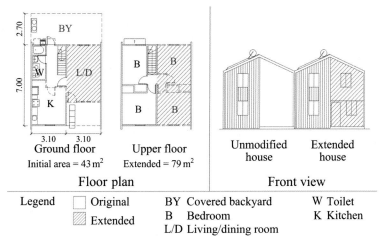

FIGURE 12.4 Villa Verde's incremental-housing design. *Credit: The authors (2020), modified from Aravena, A., & Iacobelli, A. (2012). Elemental: Incremental housing and participative design manual. Ostfildern: Hatje Cantz.*

4.1 Incremental-housing processes

From the outset of the settlement's planning, Elemental conducted design workshops with the households identified as future residents. During the first meetings, Elemental outlined the financial constraints of the project, demonstrating that the subsidies were insufficient to build houses with the size requested by the residents (Aravena, 2014). It was agreed that the incremental model, with contractors building half the house including the kitchen,

FIGURE 12.5   Examples of an unmodified house (left) and an extended house (right). *Credit: The authors (2017).*

bathroom facilities, bedrooms, and living room, would be sufficient. The residents would then complete the houses, each according to their capacities. This compromise ensured that over the longer term, each house could be expanded to meet the residents' needs and aspirations, while ensuring that the financial burdens were not concentrated in a specific period, putting at risk the financial stability of both households and funding agencies.

Elemental took on an additional role to help facilitate these future additions with the production of the habitability manual distributed to each of the households as they moved into the settlement (Arauco, 2013). The manual was a multipurpose document including tips starting from the way to maintain the house to recommendations on how to behave well within the community. It also outlined the ideal processes to double the size of the house by enclosing the area allocated for future expansion, and to complete the first and second floors to ensure structural stability. The manual detailed an efficient technique allowing the addition of walls and internal finishes and provided the required dimensions. It went so far as to provide architectural drawings of the essential building components as well as a "shopping list" of the building materials to be ordered from suppliers.

The manual also detailed the incremental process authorized by the architects, engineers, and government agencies managing the development. Thus, it was a way to guarantee that the additions were undertaken with appropriate materials capable of safely accommodating the anticipated structural loads and allowing an appropriate robustness level. The document clearly stated that any additions beyond those specified within it were considered "unauthorized" unless specific permissions had been granted by the Municipal Department of Construction, the authority responsible for assessing the associated risks to the houses, the residents, and the wider community.

Overall, Elemental's design for the Villa Verde settlement was specifically tailored to the requirements set by the clients, the regulations stipulated by Arauco and the MINVU, the physical condition of the site, and the then prevailing economic situation. Additionally, it acknowledged earthquake and tsunami risks, following the 2010 disaster, by using lightweight timber construction and developing the site away from the coast. However, it is important to note that Elemental was already well underway designing Villa Verde before the earthquake and tsunami struck. Thus, their design was not intended as a direct response to the disaster, nor as a risk-reduction initiative. Yet, the strategy to design the settlement as a low-cost social-housing project, using the half-house model to encourage self-built additions and limit the upfront costs for the developer, was certainly a timely approach for people rebuilding their lives after the disaster. Moreover, the timing of the disaster and the fact that almost half of the residents in the new settlement were affected by it allow us to identify links between the settlement's design, the built outcomes that emerged from the incremental self-build processes, and the capacity of residents to reestablish their housing.

4.2 Incremental housing in the private and public spheres

There is a great deal of evidence to suggest that the incremental-housing model was successfully embraced by the residents of Villa Verde. Possibly encouraged by the habitability manual, which demystified the building process, most residents undertook additional works to extend their houses into the front yards and backyards, although being aware that these interventions were unauthorized. Almost all houses (96%) have made steps to implement modifications outlined in the manual. The majority of the houses (85%) have completed these modifications. Nearly half of those modified houses (44%) went beyond the proposed designs with a series of modifications that included larger windows, entry porticoes, front verandas, façade decorations, carports, fences, as well as landscaping (Fig. 12.6).

The frontal additions have been complemented with significant extensions in the backyards of many houses. The process of undertaking these extensions has some significant implications at both the individual and collective levels. From the former standpoint, these additions can be viewed as positive outcomes showing that the houses enabled the residents to increase their living areas, in line with the incremental and self-help model championed by Elemental. It can be argued that these additions validate the incremental-housing strategy. Yet, from a collective standpoint, this strategy introduced some additional risks, which will be discussed in the following subsection.

FIGURE 12.6   Extensions beyond the framework set in the Elemental's habitability manual: a shopfront (left) and covered backyards (right). *Credit: The authors (2017).*

During the interviews, the vast majority of residents expressed positive attitudes toward the development and their gratitude for having received a good house. When asked what they liked about their houses, some residents responded:

I really like the house and how we adjusted it according to our needs, but we still want to modify it because the received house was very basic (female, 68);
I like it because it is my house, here I can do whatever I want. I even have a small business making "empanadas" in the oven my daughter bought (female, 72);
I like all the changes we have gradually made by ourselves and with the help of our neighbors (female, 34);
I really like the idea of progressively making adjustments and we realized that the house that looked small at the beginning can actually be large enough and comfortable once we extend it (female, 28);
I am so proud of the improvements we made in the house (male, 45).

Most residents (60%) could not specifically identify an issue they did not like about their house. Yet, the key areas of discontent focused on: concerns that the house might never be properly finished (12%), the proximity of neighbors (9%), external drainage issues (6%), internal leaking (6%), and the poor thermal performance of the house (9%).

At the household level, the residents generally believe that their dwelling needs have been catered for by the housing units. Can the same be said for the settlement's public spaces? In the initial months after the allocation of the individual housing units to the residents, the Villa Verde community organized four neighbors' committees. Over time, these committees merged with the two committees actively supported by the NGOs Urbanismo Social and ALDEA, which have assisted the community to seek funding from various governmental sources and have supported small-scale design and building programs (ALDEA, 2017; Urbanismo Social, 2016). The first completed projects included the additions of the community centers and the nursery school, which, like the houses, were only partially provided as part of the initial design and construction package. Subsequent projects have included the construction of a multipurpose room for the senior residents and benches next to the sports ground.

Aside from these organized programs, there is little evidence of significant improvements in the settlement's public spaces, despite the large area devoted to them in the project, including the cul-de-sac spaces, six other public parks surrounding the three community buildings, three playgrounds, and a basketball court. Apart from one cul-de-sac improved with additional trees, the other public spaces have seen a loss of vegetation and seating. This was because residents seek to park private vehicles closer to their houses even in spaces previously designed to be public and green. The lower importance attributed to public spaces, namely through its privatization for parking lots, contrasts with the improvements implemented in the private houses.

The Villa Verde residents have progressively worked to make improvements to their private houses and are proud of their efforts and achievements. They have also improved public buildings but have lacked the leadership and financial support to enhance the overall quality of the village—in particular the shared public spaces in cul-de-sacs and the surroundings of public buildings. The project's original visions for public green space in front of most of the houses have been subsumed by individual needs. This fact has revealed a broader situation in which private spaces and amenities are prioritized vis-à-vis public spaces or the wellbeing of neighbors and the community as a whole. This claim can be further demonstrated when examining the scope of additions made in the backyards of the properties.

4.3 Incremental housing and its drawbacks

As mentioned, most households at Villa Verde have made unauthorized extensions into the backyards. The residents of most houses (64%) have completely enclosed their backyards to add extra living spaces, whereas in a smaller cohort (21%) the backyards have been partially enclosed. These extensions represent a relatively inexpensive way to increase the size of the houses. They were usually executed with materials of low durability such as plastic, plywood, and metal sheets, which can be upgraded to more durable solutions over time. However, these types of interventions diminish the livability of the rooms inside the houses by reducing access to sunlight and ventilation for the kitchen and bathroom spaces. Hence, they add to the ongoing operational costs of the houses, which require additional lighting for extended periods. In many cases, these extensions also reduced passive solar heat gains, bringing the need for additional active heating inside the houses. Thus, these new internal spaces have compromised the livability of the houses. Besides, due to their irregular character, they may also represent an additional risk factor when an earthquake strikes.

The question remains whether the residents have considered the benefits and drawbacks of these additions and whether they were fully prepared and informed to make the associated sacrifices such as the loss of amenity and the increased heating costs. This issue is not exclusive to the Villa Verde project. Elemental's earlier housing project at Quinta Monroy has also been criticized as developing in a manner that is progressively deteriorating (Carrasco & O'Brien, 2018; Millones, 2017).

The downsides of these additions to the livability of the houses are clear. Yet, the question of whose responsibility it was to manage these shortcomings is less clear. Elemental's design appears not to have properly taken into account such a process, although it had also previously occurred at Quinta Monroy, completed in 2004. In the residents' opinion and as observed in the field, the local authorities appear to be "turning a blind eye" despite these extensions being built in Villa Verde without appropriate permits and often with low-quality construction materials.

In any case, a greater threat to human life stems from the proximity between the wooden houses. This proximity increases the risk of fire spreading from one house to another. Some residents are now keenly aware of this risk:

> I used to rent a house in the city, which was damaged by the earthquake. When we received the house in Villa Verde it was like a dream, and we moved in after six months. I am happy with my house; we expanded it little by little. Earlier this year we built an independent entrance to the first floor because the house of my father-in-law was damaged by wildfires. We are now also worried about possible fires here because the houses are made from wood (male, 34).

The fire risk is associated with two main unrelated drivers. Firstly, a fire may start within the settlement due to numerous sources, such as firewood used for heating houses burning out of control, faulty wiring, or cooking mishaps. Secondly, being located adjacent to a forested area, the settlement is exposed to wildfires. According to the National Forest

Corporation, 93% of this area is prone to wildfires and 248 wildfires were recorded in the municipality of Constitucion between 2011 and 2016 (CONAF, 2017).

5. The Villa Verde settlement examined through the Sendai Framework

This review of the incremental-housing experience at Villa Verde would not be complete without benchmarking it against the world's DRR best practice. Presently, the Sendai Framework (UNISDR, 2015), endorsed by the UN General Assembly following the 2015 Third World Conference on Disaster Risk Management, offers the most comprehensive outline in this regard. A key contribution from this framework is the acknowledgment that a wide-ranging approach to DRR is needed, encompassing the diversity of involved risks and stakeholders. Although the settlement was not exactly conceived as a disaster-response initiative, given its location in a wildfire-prone zone, it seems worthwhile to review and comment on the outcomes of the Villa Verde project based on the Sendai's four priorities.

5.1 Priority 1: Understanding disaster risk

The location of Villa Verde, high above sea level, was an effective means to manage the key tsunami risk that impacts most of the city area. Households relocated from more exposed parts of the city, ostensibly the flat areas close to downtown, are now residing in tsunami-safe houses. It is therefore undeniable that the change of location reduced some risks to the housing infrastructure. Similarly, the use of lightweight timber for constructing the frames and the cladding of the houses at Villa Verde, rather than traditional brittle-masonry systems, reduced the likelihood of structural damages and ensuing losses in lives during severe earthquake events. On the other hand, the location of Villa Verde close to forested areas—which cover 56% of all municipal land (CONAF, 2017)—exposes the settlement to wildfires. Paradoxically, the same building technology that was favored as an earthquake-mitigation measure—timber construction—is at the same time a risk factor as regards wildfires. Such risk is even exacerbated by the proximity of the houses and the unlawful self-built extensions in the backyards, which can undoubtedly facilitate the transmission of fire.

Our analyses pinpoint that, in contrast with what is proposed in Sendai's priority 1, not all the risk dimensions—"vulnerability, capacity, exposure of persons and assets, hazard characteristics and the environment" (UNISDR, 2015, p. 14)—were duly taken into account by the Villa Verde project. Indeed, the recent experience with the 2010 earthquake and tsunami disaster probably did not favor a comprehensive multi-hazard risk assessment and finally downplayed the importance of incorporating wildfire prevention and mitigation measures from the outset. Exposure and vulnerability to wildfires increased in tandem in Villa Verde. Indeed, in a report on wildfires in the municipality of Constitucion (CONAF, 2017), the National Forest Corporation highlighted the potential negative impacts of recent changes in the urban—rural interface. The scenario described in that document perfectly fits the Villa Verde settlement: populations living in higher densities in row houses built with highly inflammable materials, settled on steep slopes, a well-known wildfire-risk amplifier (CONAF, 2017, p. 9).

5.2 Priority 2: Strengthening disaster risk governance to manage disaster risk

There is evidence that the incremental-housing improvement process at Villa Verde currently lacks a strong governance system. Whereas the settlement's original consultation, design, and construction processes were once highly controlled by the developer and architect, the years after occupation have been subjected to low levels of oversight and control of the incremental development (O'Brien, Carrasco, & Dovey, 2020). Although the habitability manual stressed the need for authorizations for residents to conduct works outside those specified within it, there is no evidence to suggest that this requirement has been met. Moreover, no formal mechanism was put in place to manage these irregular additions. However, residents claimed that as of 2019 there have been some negotiations with local-government authorities for the formalization of the additions undertaken by the households.

The weak governance of the incremental process represents a risk factor, especially for a settlement built in a seismic and wildfire-prone area. It is noteworthy to recall that the Villa Verde project was developed with the collaboration of all relevant stakeholders (public and private actors, besides residents), in line with the tenets of Sendai's priority 2. Yet, this opportunity seems to have failed to consider risk governance as a key theme and to care "for the implementation of instruments relevant to disaster risk reduction" (UNISDR, 2015, p. 17). The longer-term risk implications associated with incremental development, particularly in seismic-prone countries such as Chile, have been recognized (Wakely & Riley, 2011). To effectively take DRR requirements into account, enforcing appropriate regulations and monitoring these works, often neglected parts of the process, should be embedded in incremental developments (Wainer, Ndengeyingoma, & Murray, 2016).

5.3 Priority 3: Investing in disaster risk reduction for resilience

As a reactive measure to manage wildfire risk, Arauco allocated additional land around the Villa Verde settlement since 2016 to allow for the development of a buffer zone (Figs. 12.7 and 12.8). With approximately 1500-m long and 40-m wide, this buffer surrounds both stages of the settlement. The significant amount of land finally allocated to this physical barrier, which corresponds to around 40% of the total area of Villa Verde, highlights the degree of overdue concern toward the risk of catastrophic wildfire.

In the absence of proactive measures to reduce fire risks through strategic governance models, for instance, regulating the extent of the houses' additions and the proximity between neighbors, having a buffer as a reactive structural measure in place is of utmost importance. The Villa Verde community has also been lately involved in nonstructural measures focused on risk sensitization. They established a committee that organizes fire-awareness workshops and liaises with the Arauco management and the local fire brigade. Investing in this type of risk mitigation is worthwhile but its effectiveness during a disaster might be compromised by the lack of oversight and governance regarding the factors creating the risk from the outset. Overall, the efforts toward risk prevention and reduction being carried out by Arauco and the residents in Villa Verde somehow respond to the call of Sendai's priority 3, although in a more reactive manner.

FIGURE 12.7 The first stage of Villa Verde as completed in 2013, largely embedded into the neighboring forest. *Credit: The authors (2020), based on Google. (October 3, 2013). Google Earth map of Villa Verde, Constitucion, Chile. Retrieved February 8, 2020, from http://www.earth.google.com.*

5.4 Priority 4: Enhancing disaster preparedness for effective response and to "build back better" in recovery, rehabilitation, and reconstruction

Elemental's project for Villa Verde duly took into account the earthquake and tsunami risks, which are widespread in Chile. Yet, Elemental's architects should have been aware of wildfire risk when first designing the settlement in 2010. They should also have recalled that the residents of their earlier housing estate at Quinta Monroy had extensively modified their houses in ways beyond those anticipated during the design phase. It is worth questioning whether they have been truly successful in building back better or whether their designs could have been more directly influenced by a careful analysis of the lessons learned from their earlier half-house initiative.

Moreover, for the Villa Verde residents who relocated from known earthquake- and tsunami-prone areas, wildfire might have been a new type of hazardous event to cope with. Thus, the involved stakeholders—architects, developers, and public authorities alike—should have embraced the need to prepare the community to respond to wildfires, especially in a settlement made of timber houses. Following the recommendations of Sendai's priority 4,

FIGURE 12.8   The implemented buffer zone around the two stages of Villa Verde, in 2017. *Credit: The authors (2020), based on Google. (September 9, 2017). Google Earth map of Villa Verde, Constitucion, Chile. Retrieved February 8, 2020, from http://www.earth.google.com.*

they should also have cared for duly empowering the community to "take action in anticipation of events" (UNISDR, 2015, p. 21). Indeed, the buffer zone as a mitigation measure might not be effective in case of extreme wildfires, which defy any sort of disaster response (Tedim et al., 2018). Thus, it is of prime importance that the community is equipped with a wildfire emergency plan, which includes, for instance, early-warning systems and the provision of escape routes and emergency shelters—built with fire-resistant materials.

6. Conclusion

The Villa Verde settlement has gained international attention given its promotion of incremental development via inexpensive subsidized housing that encourages residents to self-implement improvements to suit their needs and aspirations. This strategy decreases the households' initial financial investment and allows the tailored and self-paced development of houses. There is significant evidence that the half-house concept has generally been beneficial to Villa Verde's households, thanks among others to the support provided by the

Elemental designers. However, the unregulated nature of a significant part of the development, coupled with low levels of engagement on behalf of authorities in the years following the occupation, has introduced other risks that should have been acknowledged when conceptualizing the settlement.

The early design stage would be the most appropriate time to manage these issues. These would have been more easily addressed through the development of an urban plan supported with an appropriate governance model, capable to enforce the needed regulations for the sake of the community as a whole. Furthermore, by undertaking this reflective research at an early stage Elemental would have been in a better position to ensure both that the livability of the houses in the settlement was not compromised and that the risk of catastrophic fires was mitigated.

This chapter provided evidence that risks cannot ever be eliminated, even in a community in which many residents are rebounding from a disaster. The negative effects of incremental housing ended up unintentionally increasing the risk of wildfires. The experience of the Villa Verde settlement indeed brings timely lessons as regards DRR. Tighter regulations, improved governance, and strategic urban design and risk-management planning are key requirements to reduce risks in future initiatives, located in hazard-prone areas, geared toward encouraging incremental-housing and self-managed improvements.

Acknowledgments

The authors would like to thank the architect Julio Carrasco for his valuable support during data collection. We also acknowledge the important contributions from Lidia Vergara, president of the "Las Camelias" neighbors' committee in Villa Verde, the community leaders, and the residents who kindly participated in our study.

References

ALDEA. (2017). *Audience for a playground: Participatory design and construction. Villa Verde, Constitución, Chile.* Retrieved from https://www.somosaldea.org/audience-for-a-playground.

Arauco. (2013). Manual de habitabilidad: Proyecto barrio Villa Verde Constitución *[Habitability manual: Villa Verde settlement project, Constitucion].* Santiago, Chile: Arauco y Fundación Gestión Vivienda (in Spanish).

Aravena, A. (2010). Elemental—interview. *Perspecta, 42,* 85−89.

Aravena, A. (October 2014). *Alejandro Aravena: My architectural philosophy? Bring the community into the process* [Video file]. Retrieved from https://www.ted.com/talks/alejandro_aravena_my_architectural_philosophy_bring_the_community_into_the_process.

Aravena, A., & Iacobelli, A. (2012). *Elemental: Incremental housing and participative design manual.* Ostfildern: Hatje Cantz.

Burgess, R. (1992). Helping some to help themselves: Third world housing policies and development strategies. In K. Mathéy (Ed.), *Beyond self-help housing* (pp. 75−91). London: Mansell.

Carrasco, S., Ochiai, C., & Okazaki, K. (2016). A study on housing modifications in resettlement sites in Cagayan de Oro, Philippines. *Journal of Asian Architecture and Building Engineering, 15*(1), 25−32. https://doi.org/10.3130/jaabe.15.25.

Carrasco, S., & O'Brien, D. (September 2018). Incremental architecture: The case of the Quinta Monroy. In *Paper presented at the 2018 APRU sustainable cities and landscapes conference, Hong Kong.*

CONAF (Corporación Nacional Forestal). (2017). *Informe de riesgo de ocurrencia de incendios forestales en comuna Constitución [Report on the occurrence of wildfires in the municipality of Constitucion].* Retrieved from http://www.conaf.cl/wp-content/files_mf/1510588890PPCIFConstitucion.pdf (in Spanish).

Davis, I. (2013). What have we learned from 40 years' experience of disaster shelter? In D. Sanderson, & J. Burnell (Eds.), *Beyond shelter after disaster: Practice, process and possibilities* (pp. 15−34). Abingdon: Routledge.

Friedman, A. (2002). *The adaptable house.* New York, NY: McGraw-Hill.

García-Huidobro, F., Torres Torriti, D., & Tugas, N. (2008). ¡*El tiempo construye! Time builds!* Barcelona: Gustavo Gili.

Google. (October 3, 2013). *Google Earth map of Villa Verde, Constitucion, Chile.* Retrieved February 8, 2020, from http://www.earth.google.com.

Google. (September 9, 2017). *Google Earth map of Villa Verde, Constitucion, Chile.* Retrieved February 8, 2020, from http://www.earth.google.com.

Greene, M. (2017). Chile: Examples from widespread experience. In M. Nohn, & R. Goethert (Eds.), *Growing up! the search for high-density multi-story incremental housing* (pp. 75–81). Darmstadt: SIGUS-MIT & TU Darmstadt.

Groat, L. N., & Wang, D. (2013). *Architectural research methods.* Hoboken, NJ: Wiley.

Habraken, N. J. (1972). *Supports: An alternative to mass housing.* London: Architectural Press.

Junta de Andalucía. (n.d.). *Comunidad Andalucía, conjunto de viviendas en Santiago de Chile* [*Andalucia Community, housing complex in Santiago, Chile*]. Retrieved from http://www.juntadeandalucia.es/fomentoyvivienda/portal-web/web/areas/cooperacion/ArquitecturaObras/dff50f8d-a26a-11e4-9ac7-39f02d412575 (in Spanish).

Khan, T. H. (2014). *Living with transformation: Self-built housing in the city of Dhaka.* Cham: Springer.

Lizarralde, G. (2009). Post-disaster low-cost housing solutions: Learning from the poor. In G. Lizarralde, C. Johnson, & C. Davidson (Eds.), *Rebuilding after disasters: From emergency to sustainability* (pp. 25–48). New York, NY: Spon Press.

Millones, Y. (2017). La otra mitad de la Quinta Monroy [The other half of Quinta Monroy]. *Revista de Arquitectura, 22*(32), 67–72. https://doi.org/10.5354/0719-5427.2017.46147 (in Spanish).

MINVU (Ministerio de Vivienda y Urbanismo). (2017). Series estadísticas de subsidios, subsidios otorgados 1990 a Julio 2017 [*Subsidy statistics series, subsidies granted from 1990 to July 2017*]. Retrieved from http://observatoriodoc.colabora.minvu.cl (in Spanish).

MINVU (Ministerio de Vivienda y Urbanismo), Municipio de Constitución, & Arauco. (2010). Plan de Reconstrucción Sustentable PRES Constitución [*Plan for the sustainable reconstruction of Constitucion*]. Retrieved from http://minvuhistorico.minvu.cl/incjs/download.aspx?glb_cod_nodo=20100910140027&hdd_nom_archivo=Plan Maestro_PRES ConstituciónB3n_Agosto 2010.pdf (in Spanish).

Morales Martínez, R. E., Besoain Arrau, C. B., Soto Morales, A., de Carvalho, L. P., Hidalgo Pino, K. D., Fernandéz Posada, I., & Bernal Santibáñez, V. (2017). Retorno al campamento: Resistencia y melancolía en los márgenes de la ciudad formal [Back to the settlement: Resistance and melancholy on the borders of the formal city]. *Revista INVI, 32*(90), 51–75. https://doi.org/10.4067/S0718-83582017000200051 (in Spanish).

Oliver-Smith, A. (1991). Successes and failures in post-disaster resettlement. *Disasters, 15*(1), 12–23. https://doi.org/10.1111/j.1467-7717.1991.tb00423.x.

O'Brien, D., & Ahmed, I. (2012). Stage two and beyond: Improving residents' capacity to modify reconstruction agency housing. In S. Mizokami, R. Kakimoto, & F. Yamada (Eds.), *Proceedings of the 8th Annual Conference of the International Institute for Infrastructure Renewal and Reconstruction on Disaster Management, Kumamoto, Japan* (pp. 309–317). Retrieved from https://iiirr.ucalgary.ca/files/iiirr/B3-4_.pdf.

O'Brien, D., Carrasco, S., & Dovey, K. (2020). Incremental housing: Harnessing informality at Villa Verde. *Archnet-IJAR.* https://doi.org/10.1108/ARCH-10-2019-0237. ahead-of-print.

Rodríguez, A., & Sugranyes, A. (Eds.). (2005). Los con techo: Un desafío para la política de vivienda social [*Those with a roof: A challenge for social-housing policy*]. Santiago, Chile: Ediciones Sur (in Spanish).

Salvi Del Pero, A. (2016). *Housing policy in Chile: A case study on two housing programmes for low-income households.* https://doi.org/10.1787/5jm2hzbnqq33-en. OECD Social, Employment and Migration Working Paper No. 173.

Schilderman, T. (2004). Adapting traditional shelter for disaster mitigation and reconstruction: Experiences with community-based approaches. *Building Research & Information, 32*(5), 414–426. https://doi.org/10.1080/0961321042000250979.

Tedim, F., Leone, V., Amraoui, M., Bouillon, C., Coughlan, M., Delogu, G., … Xanthopoulos, G. (2018). Defining extreme wildfire events: Difficulties, challenges, and impacts. *Fire, 1*(9). https://doi.org/10.3390/fire1010009.

Tipple, A. G. (2000). *Extending themselves: User-initiated transformations of government-built housing in developing countries.* Liverpool: Liverpool University Press.

Turner, J. F. (1968). The squatter settlement: An architecture that works. *Architectural Design, 38,* 355–360.

Turner, J. F., & Fichter, R. (Eds.). (1972). *Freedom to build: Dweller control of the housing process.* New York, NY: Macmillan.

UN-Habitat. (2011). *Affordable land and housing in Latin America and the Caribbean.* Nairobi: UN-Habitat.

UNISDR (United Nations Office for Disaster Risk Reduction). (2015). *Sendai Framework for Disaster Risk Reduction 2015–2030*. Retrieved from https://www.preventionweb.net/files/43291_sendaiframeworkfordrren.pdf.

Urbanismo Social. (2016). La convivencia en el barrio es una preocupación constante para las comunidades *[Coexistence in the neighborhood is a permanent concern for the communities]*. Retrieved from http://www.urbanismosocial.cl/la-convivencia-en-el-barrio-es-una-preocupacion-constante-para-las-comunidades (in Spanish).

Van Hoogstraten, D. (2000). Between structure and form: Habraken and the alternative to mass housing. In K. Bosma, D. V. Hoogstraten, & M. Vos (Eds.), *Housing for the millions: John Habraken and the SAR 1960–2000* (pp. 87–142). Rotterdam: NAi Publishers.

Wainer, L., Ndengeyingoma, B., & Murray, S. (2016). *Incremental housing, and other design principles for low-cost housing*. In *International growth centre's report C-38400-RWA-1*. Retrieved from https://www.theigc.org/wp-content/uploads/2016/11/Wainer-et-al-2016-final-report.pdf.

Wakely, P., & Riley, E. (2011). *The case for incremental housing*. Cities Alliance Policy Research and Working Papers Series No. 1. Retrieved from http://documents.worldbank.org/curated/en/883891468150578554/The-case-for-incremental-housing.

Yin, R. K. (2017). *Case study research and applications: Design and methods*. Los Angeles, CA: Sage.

Zeisel, J. (1984). *Inquiry by design: Tools for environment-behavior research*. Cambridge, MA: Cambridge University Press.

Climate action zones: A clustering methodology for resilient spatial planning in climate uncertainty

Kelly Leilani Main[1], Miho Mazereeuw[1], Fadi Masoud[2], Jia Lu[2], Aditya Barve[1], Mayank Ojha[1], Chetan Krishna[1]

[1]Massachusetts Institute of Technology, Urban Risk Lab, Cambridge, MA, United States;
[2]University of Toronto, Centre for Landscape Research, Toronto, Ontario, Canada

1. Introduction

Rapidly changing environmental conditions in coastal areas warrant a new approach to resilient land-use planning and urban design. Although equitable resilience (Matin, Forrester, & Ensor, 2018) necessitates the consideration of socioeconomic features when envisaging land-use plans and future development, we argue that underlying environmental conditions should be the primary departure point for understanding the use value of particular land parcels. This is because the environmental context of each place necessarily determines the sustainability and survivability of any climate-adaptive or resilient programs and design interventions. With environmental considerations as a foundation, socioeconomic variables can be added as supplementary layers to guide policy and design decisions.

To explore this concept, we examined one of the areas of the United States that are most susceptible to climate-change impacts: Southeastern Florida. Three-fourths of Florida's population reside in coastal counties that generate 79% of the state's total economic activity. These counties combined represent a built environment and infrastructure whose replacement value by 2030 is projected to be USD3 trillion (FOCC, 2010). Broward County, located just north of Miami-Dade, will face a property loss of USD24.4 billion—17.5% of the total tax base—due to a potential 1-m sea-level-rise scenario. Moreover, while the slow-onset risks of sea-level rise already threaten built assets, the increasing severity and intensity of extreme weather events such as hurricanes will bring heavy rainfall and storm-surge scenarios to other areas not already facing chronic "sunny-day" flooding and king tides. Additionally, Southeastern Florida largely comprises porous limestone bedrock, resulting in saltwater intrusion that threatens the viability of infrastructure and limits opportunities for water retention in low-lying areas. To address the numerous risks to property and infrastructure in this region, planners, policymakers, and designers must consider the complex interrelations of the natural and built environment by developing new approaches to land-use planning. In areas where climate-change impacts are still uncertain, we argue that resilient development typologies based on socioecological resilience frameworks can help guide innovation in architecture and design.

This chapter explores an experimental approach for delineating climate action zones (CAZs) based on the unique environmental-risk profile of Broward County. Section 2 describes the etymology of resilience, adaptation, transformation, and uncertainty to articulate why a socioecological resilience approach is important for dealing with uncertainty in the built environment. Section 3 critiques current methods of land-use planning to highlight the need for a new approach that is grounded in environmental processes at larger scales. Section 4 introduces the nonhierarchical k-means/k-prototypes clustering methodology to analyze similarities in large datasets. Section 5 presents our results, the eight distinct CAZs, and

Section 6 provides strategies for addressing what we define as "zones of uncertainty"—areas in which innovations in planning and design methods can use socioecological resilience frameworks as inspiration for uncertain futures. As expressed in the limitations and the conclusions sections, this methodology is not yet a panacea for addressing resilience and uncertainty associated with climate change. Rather, we hope that this novel technique will help local authorities incorporate environmentally sound and data-driven approaches to transform traditional land-use planning and increase the resilience of their communities.

2. Resilience, adaptation, transformation, and uncertainty

The etymology of resilience crosses multiple disciplines including mechanical engineering, psychology, and ecology. A common conceptualization of resilience as it applies to the built environment derives from engineering, which describes a buffer from failure (Adger, 2000) or an ability to "bounce back" to an original functioning state (Alexander, 2013). In contrast, ecological resilience refers to a system's ability to absorb change or disturbance and adapt (Holling, 1973). Adger, Hughes, Folke, Carpenter, and Rockstrom (2005, p. 1036) defined resilience as the "capacity of linked social-ecological systems to absorb recurrent disturbances such as hurricanes or floods so as to retain essential structures, processes, and feedbacks." Since the development of the socioecological resilience literature, the term resilience has grown to become a catchall concept within the climate-change adaptation and disaster-preparedness fields (Biggs, Schlüter, & Schoon, 2015; Folke, 2006; McGinnis & Ostrom, 2014; Pike, Dawley, & Tomaney, 2010).

As decisions about adapting to climate change and being resilient to its impacts involve inevitable trade-offs (Schoon et al., 2015), institutions and policies may need to adjust their strategies to give the opportunities to communities to be more resilient to such impacts. However, incremental adaptation is not always viable for developing long-term resilience to systems' collapse (Handmer & Dovers, 1996). Instead, a complete transformation of existing systems may be necessary. Whereas there is some debate as to whether adaptation necessarily includes transformation, transformation refers to a fundamental alteration of the nature of a system in response to untenable or undesirable conditions (Nelson, Adger, & Brown, 2007). Through transformation, new activity pathways are possible.

Transformation may become increasingly important as climate scientists estimate that it is not possible to "reverse" climate change; for, even if greenhouse gas emissions are halted, the effects of climate change will continue for centuries (Gray, 2007; Holling et al., 2004; Steffen et al., 2015). The possibilities of future destabilization are marked with uncertainty, "a cognitive state of incomplete knowledge that results from a lack of information and/or from disagreement about what is known or even knowable" (Pachauri & Meyer, 2014, p. 128). Human inaction on reducing greenhouse gas emissions has irreparably transformed the biosphere in such a way that we now face fundamental uncertainties about the future (Hilborn, 1987; Peterson et al., 1997). Although uncertainty has entered the literature regarding climate projections (Deser, Phillips, Bourdette, & Teng, 2012; Jones, 2000; Latif, 2011), sea-level rise (Rahmstorf, 2007), and human behaviors and social responses to climate change (Moser, 2005), there remains a dearth of examples of how to manage uncertainty through policy and design projects in the built environment.

One of the ways the climate-adaptation practice has addressed uncertainty is through the process of adaptive management, which argues that because climate-change impacts are difficult to predict, flexible strategies are essential for managing the potential stresses and opportunities (Folke, 2006; Hallegatte, 2009; Lee, 1999; Nelson, 2011). This process includes adaptive governance, which requires institutional structures to be flexible across a variety of scales and with reasonable degrees of autonomy (Folke, Hahn, Olsson, & Norberg, 2005). Although strategies for adaptive management exist (Hallegatte, 2009), even a flexible process of adaptive management will encounter its own barriers (Stankey, Clark, & Bormann, 2005). One such barrier may be institutions themselves. Although they often exist to reduce economic, social, and political transactional costs, institutions can also be inflexible regarding the speed at which environmental and social changes may occur (Barnett et al., 2015). We argue that there is a need to transform land-use planning methods to address climate uncertainty while improving the resilience of the built environment.

3. Adaptive management and the transformation of traditional land-use planning methods

As alluded earlier, institutions are not necessarily well suited for adapting to the uncertainties of climate change. This is partially due to path dependency, the durability or persistence of a decision made based on past conditions that leads to undesirable outcomes and is costly to change (Margolis & Liebowitz, 1995; Sorenson & Hess, 2015). Path dependency in land-use planning can be attributed to two primary reasons. First, traditional urban planning relies on a didactic legal practice of regulating land and its use. The effectiveness of this normative planning model is intertwined with the idea that landscape is static. As a result, static land-use plans generated using conventional processes have a time horizon of 20 to 30 years, after which they are updated for another static period. Such land-use plans rarely use underlying geomorphological features (such as water) to determine the use and regulation of land beyond the distinction between what is buildable area versus what is not. Over time, investments in different land uses (residential, industrial, commercial) lead to feedback loops related to existing uses (generating economic agglomeration, neighborhood character, nuisance law, etc.) and as a result, future land-use planning often relies on existing plans as the foundation for subsequent ones, even if environmental conditions have drastically changed. Although incremental adaptation may occur, land-use regime changes are relatively minor, as shown by comparing two Broward County land-use plans from roughly 30 years apart in Fig. 13.1.

The second limitation in current modes of land-use planning is that it is defined by arbitrary political boundaries. Though adaptation and resilience approaches must be multiscalar because what is resilient at one scale is not necessarily resilient at another (Adger, Arnell, & Tompkins, 2005), land-use planning is proceeded most often at the level of the municipality. Although Broward County is unique for having a county-level land-use plan, municipalities participate in the planning process and therefore the county is more of a compendium of local land-use plans than a collective strategic vision. This contrasts with the ecological processes that shape the landscape, especially the hydrological ones, which transcend these boundaries.

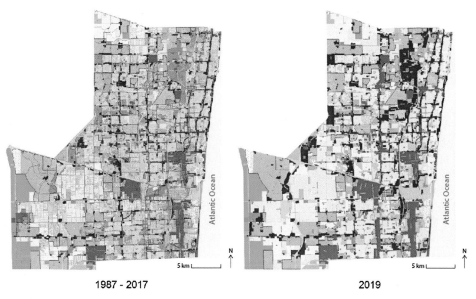

1987 - 2017   2019

FIGURE 13.1   Residence, represented in yellow, remains the most ubiquitous land use in Broward County across the decades (red indicates commercial mixed use, purple industry, gray transportation, and green recreation and open space). *Credit: Broward County Land Use Map Adopted—December 3, 2019.*

The current development paradigm of planning static "end-state" building regulations is based on rigid codifications at hyper-localized scales. To be more resilient, it should give way to flexible and ecologically sensitive development codes attuned to locally dynamic geomorphological processes. This move will require up-scaling land-use planning to wider geographical spheres and more interagency coordination among different administrative areas (Bedsworth & Hanak, 2010; Bulkeley & Castán Broto, 2013; Zimmerman & Faris, 2011). A broader approach to regional land-use planning based on environmental conditions would also lead to more appropriate plans at local scales. Geospatial information and data-processing tools can help spatial planners identify where environmentally appropriate actions can be taken, rather than relying completely on incremental adaptation of existing land-use plans at municipal scales. The following section describes how we used geospatial data and machine learning to identify categorical clusters in Broward County.

4. A nonhierarchical clustering approach using *k*-means/*k*-prototypes

A new paradigm of land-use planning will require instruments capable of regulating ecological processes—tools and methods that can negotiate uncertain ecological flux with the regulatory authority of land-use planning. The use of models—in this case a clustering algorithm intended to identify geomorphologically similar areas—can help to detect which adaptation actions are more appropriate and also "bound the range of our uncertainty" (Peterson et al., 1997, p. 7). By clustering a region's geomorphology into clusters derived from environmental features rather than municipal boundaries, this approach can break down the path-dependency restrictions of existing land-use-planning methods. It can also help residents, and local and county-level officials to collectively understand how climate impacts will cross municipal boundaries and affect the county as a whole. This section describes an experimental methodological model for identifying zones for different types of adaptation actions.

To begin, we analyzed the landscape of Broward County using geospatial datasets with ecomorphological attributes such as soil permeability and seasonal water-table heights, salt-water inundation, storm-surge heights, and ground elevation (see Table 13.1). In our case, we worked at the county scale, but it is important to note that this method could be applied at any scale and should not rely on administrative boundaries. Although most of the used datasets are based purely on environmental features, it is important to highlight that we also used data from the Federal Emergency Management Agency (FEMA) as part of our analysis. Although this dataset is technically representative of flood-insurance rates, it can also be used for evaluating flooding risk based on their calculated annual percentage chance of flooding derived from historical data.

We used a nonhierarchical k-means/k-prototypes clustering algorithm to identify eight landform classes based on similar properties. The k-means conceptual clustering algorithm (MacQueen, 1967) partitions data into disjoint sets based on similarities against cluster means. It is well known for its efficiency in the face of very large datasets and has been used for geospatial applications such as hazard prediction (Gorsevski, Gessler, & Jankowski, 2003) and landform classification (Burrough, Wilson, Van Gaans, & Hansen, 2001). Classification of the landmass using geo-referenced data to determine discrete types and their corresponding boundaries has been widely explored for climate zones (Netzel & Stepinski, 2016). Similar methodologies have been applied to delineate management zones for urban flood risk (Xu, Ma, Lian, Xu, & Chaima, 2018), categorization for land-use alternatives (Smith & Saito, 2001), and site-management in precision agriculture for coastal saline land (Li, Shi, Li, & Li, 2007).

Given the mixed numerical and categorical nature of the descriptor data, we employed a variant of the k-means, the k-prototypes algorithm (Huang, 1997). Through this approach it is possible to measure object similarity against a prototype containing both numerical and categorical data. The algorithm is summarized as follows:

— Select k initial prototypes from the dataset. There are several methods of allocating the initial set of k prototypes (see, for instance, Ji et al., 2015). We used the farthest-point heuristic-based initialization discussed in He (2006).
— Allocate each object in the data to one of k clusters based on determined similarity to the k prototypes. The similarity of an object to a prototype is determined by a construct of the Euclidean distance between numerical attributes and a weighted "difference" of the categorical attributes.
— Update the prototype of the cluster after each allocation, that is, reinitialize the prototypes. After allocating all objects, retest the similarity of objects against the k prototypes. If an object is found to belong to another cluster based on similarity to the updated prototypes, reallocate the object and update both clusters.
— Repeat the earlier step until no further reallocations occur.

TABLE 13.1 Datasets used for clustering.

Dataset	Raw-data parameters	Input parameters for clustering	Description	Data source
Soil data	Seasonal high-water-table range	Normalized soil-saturation index (numerical)	Ground saturation near surface, factoring water-table depth and duration for which high-water-table conditions persist	South Florida Water Management District, Natural Soil Landscape position 1999
Soil data	Seasonal high-water-table range	Normalized duration-variability index (numerical)	Variation in seasonal high-water-table duration; values are such: highly variable = 100; precise value (e.g., 4 months) = 0; range values are computed, e.g., 4–7 months: (7−4) *100/12 = 25.	South Florida Water Management District, Natural Soil Landscape position 1999
FEMA flood-hazard zones	Flood zone	Flood-zone index (categorical)	**VE**: High-risk coastal area: coastal areas with a 1% or greater chance of flooding and an additional hazard associated with storm waves **AO**: River or stream flood-hazard areas, and areas with a 1% or greater chance of shallow flooding each year, usually in the form of sheet flow, with an average depth ranging from 0.3 to 1 m **AE**: The floodplain where BFE in relation to NGVD is provided **AH**: Areas with a 1% annual chance of shallow flooding, usually in the form of a pond, with an average depth ranging from 0.3 to 1 m. These areas have a 26% chance of flooding over the life of a 30-year mortgage **X**: Area of moderate flood hazard, usually the area between the limits of the 100-year and 500-year floods	FEMA national flood-hazard layer data
SLOSH model	Storm-surge category	Storm-surge index (numerical)	Category of lowest-intensity storm that can affect an area through storm surge. Surge zones were created using a surge-modeling application for the Florida statewide Regional evacuation Update study for each category of storm (1–5).	National storm surge hazard maps, version 2, NOAA

(*Continued*)

TABLE 13.1 Datasets used for clustering.—cont'd

Dataset	Raw-data parameters	Input parameters for clustering	Description	Data source
Digital-elevation model	Ground elevation, static-base flood elevation (from FEMA flood-hazard zones)	Flood-height index (numerical)	Normalized index computed from the difference of static BFE and ground elevation. Used in conjunction with flood-zone index and storm-surge index to indicate the vulnerability to flooding as a factor of ground elevation and the 500-year flood line	ASTER GDEM version 2, maintained by the NASA EOSDIS LP DAAC

Key: ASTER GDEM, Advanced Spaceborne Thermal Emission and Reflection Radiometer Global Digital Elevation Model; *BFE*, Base Flood Elevation; *NASA EOSDIS LP DAAC*, NASA Earth Observing System Data and Information System Land Process Distributed Active Archive Centre; *NGVD*, National Geodetic Vertical Datum; *NOAA*, National Oceanic and Atmospheric Administration; *SLOSH*, Sea, Lake and Overland Surge from Hurricanes.

The "error" E, or cost of each allocation, can be split into two distinct components E_r and E_c, numerical and categorical respectively, and each is the sum of the similarity measure or "distance" of each object in a cluster to its respective prototype or sum of the intra-cluster distances. The total error is this sum further added across all clusters. This algorithm is applied repeatedly, varying k until a distinct number of clusters is formed, which are amenable to interpretation and categorize distinctions, yet not too large to become intractably complex. When k equals the number of objects in the dataset, the error is zero, and when k is 1, the error is maximum. The error is a decreasing monotonic function of k. Several methods exist for evaluating the clustering for a particular k based on variance and information measures of the partitioning where we do not have the "ground truth" of correct cluster assignments (Maulik & Bandyopadhyay, 2002). The choice of the final k is often somewhat subjective and can be justified by the information derived from such appropriate metrics. In our case, we chose k based on the "elbow method" (Ketchen & Shook, 1996), according to conventional practice, selecting k near the inflection point in the $k-E$ graph (Fig. 13.2).

As the number of clusters increases, the difference between clusters dramatically decreases. Thus, the goal is to select a number of clusters that are differentiated enough to be able to discern them from one another and yet not too numerous. In fact, if there are too many clusters, the ensuing implementing strategies become restrictively complex. After selecting an appropriate number of clusters, we conducted an analysis of the qualities of each cluster and derived categorical definitions based on their profiles (Fig. 13.3). The results of these categories are illuminated in the results section.

5. Identification of eight climate action zones

Using the abovementioned method, we identified eight distinct clusters (Fig. 13.4). Based on each cluster's unique profile, we grouped them into three categories: high risk, low risk, and zones of uncertainty. These designations can function as a starting point for adaptation policy, planning, and design decisions. For example, high-risk zones may be sites for

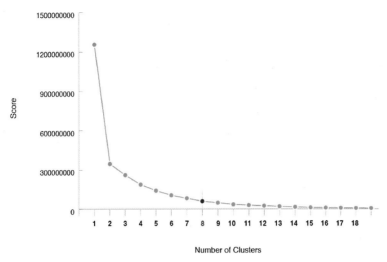

FIGURE 13.2 Plot showing score for *k* clusters (*E*) versus number of clusters (*k*), indicating the selection of *k* = 8. *Credit: The authors.*

downzoning or managed-retreat programs, and low-risk zones may be more suitable for densification and future development. Zones of uncertainty include areas that have either a lack of available data or variability in the data that makes it difficult to decide on a precise path forward, but provides an opportunity for innovation in planning and design practices.

5.1 High-risk zones

5.1.1 Low-lying urban landscape (Cluster 1)

This is predominantly an urban, inland cluster where most parcels are located in FEMA flood zones that require the purchasing of flood insurance through the National Flood Insurance Program (NFIP). Because most parcels are urban, there is little information on soil type concerning permeability and saturation. Home elevation is an example of a strategy that may provide opportunities for residents to remain in the short term, but as risks increase into the future, these areas will need to plan for potential shifts in land-use planning.

5.1.2 Uninsured inland areas (Cluster 2)

This cluster is categorized by high-saturation soils, including marl and rocky soils, muck depressions, Everglades peat, and tidal areas, with water tables ranging from 0.3 m below to 0.6 m above ground level. Less than 2% of homes require flood insurance, but many parcels can be affected during a category-4 hurricane. The fact that there is already high soil saturation means that there are limited water-retention options available onsite, making these areas particularly vulnerable. Proper water-management strategies will be essential to ensuring that this zone can be developed despite being at risk of pluvial flooding.

5.1.3 Coastal wetlands (Cluster 3)

This cluster represents the most vulnerable tidal area, even during minor storms. Almost all parcels are located in FEMA flood zone AE (that is, with 1% chance of annual flooding)

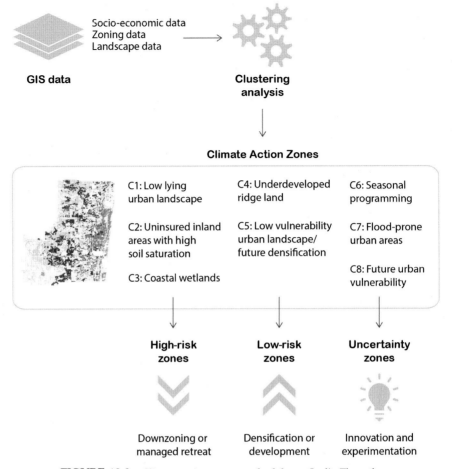

FIGURE 13.3   Climate action zones methodology. *Credit: The authors.*

and thus require flood insurance through NFIP. Adaptation actions and transitions to open space are likely to be necessary in the future.

5.2 Low-risk zones

5.2.1 Underdeveloped ridge land (Cluster 4)

Parcels in Cluster 4 generally have the lowest soil saturation, with most falling within the central ridge, knoll, and flatwood soils category. Ridge and knoll soil-type parcels generally are located at a higher elevation, and as they are not categorized as urban, they have relatively less development on them. Low flood risk means that less than 2% of homes require flood insurance, but some parcels can be affected by storm surge during category-3 or 4 hurricanes. Low storm-surge risk means that these zones may be appropriate sites for strategic up-zoning and densification. Low soil saturation means water retention options are viable onsite.

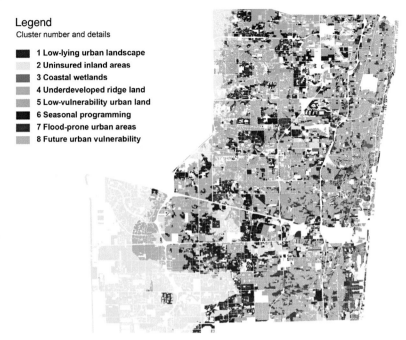

Legend

Cluster number and details

■ **1 Low-lying urban landscape**
□ **2 Uninsured inland areas**
■ **3 Coastal wetlands**
▨ **4 Underdeveloped ridge land**
▨ **5 Low-vulnerability urban land**
■ **6 Seasonal programming**
■ **7 Flood-prone urban areas**
▨ **8 Future urban vulnerability**

FIGURE 13.4   A visualization of the eight climate action zone clusters as an output of the *k*-means approach. *Credit: The authors.*

5.2.2 *Low-vulnerability urban landscape/future densification (Cluster 5)*

No parcels in Cluster 5 are in a storm-surge zone. Therefore, flood insurance is not required for any homes located here. However, most parcels correspond to urban areas (three-quarters of all urban lands), which means that soil-type and water-table data is highly variable. With low susceptibility to environmental hazards and high urban build-up, this area can most likely see continued urbanization and densification in the future. Planners may focus on encouraging development in these areas.

5.3 Zones of uncertainty

5.3.1 *Seasonal programming (Cluster 6)*

Cluster 6 is a zone with no insurance required, but it has many parcels that can be at risk during category-3 and 4 storms. One of its features is the exclusive presence of flat soils, with a highly variable water table of -0.3 to 0 m. Because of this variability, seasonal programming or innovative water-retention strategies may be an interesting way to manage flood risk in this area.

5.3.2 *Flood-prone urban areas (Cluster 7)*

Cluster 7 is almost entirely an urban cluster, where most parcels require flood insurance due to their location in FEMA zone AE. Parcels here may flood in times of increasing runoff, storm surge, or experience "sunny-day" flooding. All parcels in this cluster can be impacted by a category-5 storm or lower. However, their urban nature means that local context should help guide an overall strategy. Policymakers and planners may consider establishing "no-build zones" or buyout programs for disaster response. Critical infrastructure should be relocated if possible, and recreational use is encouraged.

5.3.3 Future urban vulnerability (Cluster 8)

Cluster 8 is entirely an urban cluster, in which less than 2% of homes require flood insurance under the FEMA NFIP program. Most parcels are not at immediate risk of flooding. However, exposure to future flood hazards will likely increase as sea levels rise due to high tides and storm surge, because all parcels will be impacted by a category-5 storm-surge event. These areas may face significant environmental changes over the next 30 to 50 years and planners may decide to develop long-term visioning and innovative policies to encourage strategic adaptation. Because the area is quite large, it may also allow for some innovative architectural and urban-design strategies.

6. Strategies for resilient design in zones of uncertainty

The results outlined earlier display an innovative way for planners and designers to plan for flood risk in different areas of Broward County. Although some high-risk zones may need to decrease in density over time through relocation programs and incentives, other areas are less exposed to flood hazards and are thus more suitable for densification, hosting critical services, and attracting investment. In areas of uncertainty, there may be opportunities for experimentation where climate risks are highly variable, or where repercussions are difficult to quantify or predict. The results from the CAZ methodology identify three clusters—Seasonal programming (Cluster 6), Flood-prone urban areas (Cluster 7), and Future urban vulnerability (Cluster 8)—in which the implications for adaptation action are uncertain. Drawing from the socioecological resilience literature, the following subsections describe some strategies that may be useful for urban planners and designers in experimenting in zones of uncertainty. In these areas, the principles of modularity of connectivity and functional diversity (Biggs et al., 2015) can guide novel and climate-adaptive design and policy strategies at multiple scales.

6.1 Modular connectivity

Connectivity is defined here as the structure and strength of the interactions between resources, species, or actors within a system (Dakos et al., 2015). Accordingly, modular connectivity describes the different combinations of functions of CAZs across the county because each zone has a unique function within a connected socioecological landscape. Modular connectivity also describes how increasing the compartmentalization of heterogeneous system components (i.e., lakes, which are functionally organized as independent clusters that are very loosely connected to each other) can buffer the spread of disturbances and adjust more gradually to them (Scheffer et al., 2012).

At the scale of the county, the levels of risk are defined by distinct geomorphological zones that have unique characteristics and systems. Thus, the county can benefit from the inherent modularity of the landscape by determining the level of compartmentalization and connectivity between zones. In contrast, current administrative boundaries ignore underlying landscape qualities and result in similar land-use patterns across municipalities with different environmental opportunities and constraints. As cities can only regulate their land use, and thus benefit from their own individual tax base, no single municipality will likely opt for buyout or retreat programs out of highly hazard-prone areas due to inertia and the unwillingness to forgo sunk costs. However, at the county scale, in which cities share critical infrastructures such as sewage and waste-water treatment plants, if a system fails in one municipality, it can have cascading effects for its neighbors. Thus, some areas of a county should favor mitigating risks by reducing density while increasing investment in areas less exposed to hazards.

At the parcel level, modularity can increase the resilience of individual households and neighborhoods in high-risk areas simultaneously. For example, single buildings that filter rainwater to potable standards (Hawken, 2017) can be cut off from municipal water systems and groundwater wells to prevent the spread of contamination. By examining the potential trajectories of disruptions, designing for modularity can mitigate the spread of disturbances. These "buffer" areas may also provide opportunities for the development of closed-loop systems, like district energy or recreational space, but this would require new strategies for municipal governance and partnerships across scales.

6.2 Functional diversity

A second guiding principle for resilience building within systems ecology is functional diversity, which emphasizes the distribution of system components rather than simply their diversity in numbers (Walker, Kinzig, & Langridge, 1999). Diversity and redundancy within a group of system components with similar functions can increase resilience and provide options for responding to disturbance and surprise (Kotschy, Biggs, Daw, Folke, & West, 2015). Functional diversity uses the same risk-mitigation principle used by diversified stock portfolios to increase the resilience of cities, landscapes, and regions to unpredictable climate change disturbances. This approach emphasizes that uncertainty is inherent to complex systems and that each intervention might result in externalities (Homer-Dixon et al., 2015; Walker & Salt, 2012). Unlike traditional policy prescriptions, a diversity of strategies in the short term can generate more adaptation options in the long term, minimizing the risks of path dependency, lock-in, and maladaptation.

At the parcel level, designers and planners can encourage that public spaces have multiple functions, such as below-grade basketball courts also serving as water-retention ponds. At the neighborhood level, experimentations can allow comparison of results over time. To illustrate, one neighborhood might employ bioswales in combination with rain gardens, and another a series of rain barrels and green roofs. Over time, the most successful strategies will emerge as an option to be scaled up. Similarly, within CAZ Cluster 6 (Seasonal variability), different seasonal-use patterns could mean altered public-transportation routes in one area, and in another it may mean seasonal housing for tourists instead of a year-round residential community. A wide variety of options could encourage both multiple solutions that are desirable for property owners and new county-wide strategies, especially in the CAZs with high variability and uncertainty.

7. Limitations

Although CAZs provide opportunities for addressing environmental hazards and dealing with uncertainty, this methodology does face limitations. The first one is inherent to its technical approach. For instance, in our case, the selection of eight clusters was a subjective choice based on the results of a clustering algorithm. Thus, other choices near the inflection point on the error curve (Fig. 13.2) may be equally suitable. The characteristics of each cluster will inherently show similarities as the number of clusters is reduced. Having a higher number of clusters may, on the other hand, yield more distinct clusters with more granular dissimilarities. A rigorous approach when identifying clusters, particularly when considering policy interventions, would be to develop multiple sets of clusters and examine how salient the classifications would be for the proposed actions. Such an approach would go beyond using other common metrics such as the Calinski-Harabasz criterion or David-Bouldin criterion to root evaluation in on-ground conditions. Although we did not follow this approach in this chapter, this is an important consideration when testing and applying such a method in the future.

A second challenge is limitations with data, as it may be difficult to replicate this method in other regions that lack substantial ecomorphological datasets. As the availability of geospatial data increases, hopefully this methodology can be tested in other contexts. The third limitation concerns the complexity of any intervention based on this method due to path dependency and institutional lock-in. Transforming current static land-use planning practices and building codes will be challenging without institutional reforms that allow for nimble and flexible planning. Thus, even if this method proposes a conceptually suitable basis for resilient land-use planning, it will be politically difficult to transform land-use planning procedures to meet the recommendations of the findings. Fourth, within zones of uncertainty, seeding diverse solutions for future propagation is often costlier, as efficiency is decreased when functional diversity and redundancy are increased (Kotschy et al. 2015; Walker et al., 1999). Facilitating the implementation of functional diversity and modularity of connectivity across parcels within a municipality and between municipalities at the scale of the county can be an extremely time- and resource-intensive activity. However, the initial costs of an innovation period in the short term may be minimal when viewed through the longer time horizon needed to build resilience to climate-change impacts.

Lastly, whereas CAZs present a novel way to approach planning in coastal areas, decisions about community development cannot be derived solely from environmental data. However, it is also true that the existing approach to land-use planning falls short in terms of preparing coastal communities for the environmental risks associated with climate change. Thus, the CAZ methodology and the results of its application above show an example of how planners can use environmental data to inform strategic adaptation decisions across multiple scales. Once this foundation has been laid, land-use planning institutions can take specific decisions about which adaptation strategies to use, while choosing the best timing for activating community-engagement processes. Indeed, by allowing stakeholders to participate in the decision-making about their future in multiple scales, these processes complement with socioeconomic inputs the geomorphological approach of the CAZ method. Likewise, planning and design officials should continue to use these dynamic data inputs to inform adaptive management in land-use planning into the future as new data becomes available and environmental conditions change.

8. Conclusions and future research

In this chapter, we illustrated an experimental methodology for using geospatial and flood-risk data to develop a classification of CAZs in a region subject to climate-change related disturbances. The resulting CAZ model provides planners and policymakers with a starting point from which to build resilience to such disturbances through adaptive land-use planning rooted in ecomorphological attributes. Rather than relying on existing land-use plans as the basis for future spatial planning, CAZs provide a different framework for understanding future hazards and proposing adapted development strategies for each parcel of land (Fig. 13.5).

Distinguishing between high-risk, low-risk, and uncertainty-oriented CAZs, this approach focuses on strategies that are dynamic and work across differing space and time scales. We utilize ecological-resilience characteristics such as modularity of connectivity and functional diversity to inform innovative design in different zones. The methodology illustrated in this chapter may provide new insight for socioecological resilience principles informed through data science and environmental conditions. Future research is needed to test this methodology with additional datasets such as sea-level rise projections and the comparison of different clustering categories. Furthermore, to make this approach actionable, it is important that the methodology is easily accessible and understandable. For this to be possible, environmental literacy must be increased among impacted communities.

We are continuing our experimentations with a dynamic digital web-cartographic tool that provides a platform for urban planners and decision-makers, and allows them to visualize different geospatial and demographic layers, climate action zones, and associated design ideas. The next steps in this research include experimenting with new clustering methodologies, replicating the analysis in other regions, and developing implementable design strategies utilizing the resilience principles illustrated here. This methodology is a preliminary exploration of spatial planning and design that combines managing risks with opportunities to innovate in the face of the unknown.

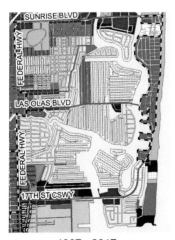 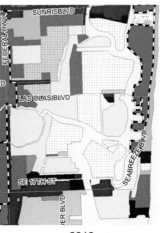 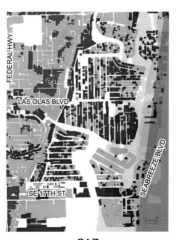

| 1987 - 2017 | 2019 | CAZ |

FIGURE 13.5 Climate action zones inform radically different uses in a selected area of Broward County compared to traditional land-use maps. *Credit: The authors.*

Acknowledgments

The author would like to thank the Broward County's Environmental Protection and Growth Management Department for their support to this research.

References

Adger, W. N. (2000). Social and ecological resilience: Are they related? *Progress in Human Geography, 24*(3), 347–364. https://doi.org/10.1191/030913200701540465.

Adger, W. N., Arnell, N. W., & Tompkins, E. L. (2005). Successful adaptation to climate change across scales. *Global Environmental Change, 15*(2), 77–86. https://doi.org/10.1016/j.gloenvcha.2004.12.005.

Adger, W. N., Hughes, T. P., Folke, C., Carpenter, S. R., & Rockstrom, J. (2005). Social-ecological resilience to coastal disasters. *Science, 309*(5737), 1036–1039. https://doi.org/10.1126/science.1112122.

Alexander, D. E. (2013). Resilience and disaster risk reduction: An etymological journey. *Natural Hazards and Earth System Sciences, 13*(11), 2707–2716. https://doi.org/10.5194/nhess-13-2707-2013.

Barnett, J., Evans, L., Gross, C., Kiem, A., Kingsford, R., Palutikof, J., … Smithers, S. (2015). From barriers to limits to climate change adaptation: Path dependency and the speed of change. *Ecology and Society, 20*(3). https://doi.org/10.5751/ES-07698-200305.

Bedsworth, L. W., & Hanak, E. (2010). Adaptation to climate change: A review of challenges and tradeoffs in six areas. *Journal of the American Planning Association, 76*(4), 477–495. https://doi.org/10.1080/01944363.2010.502047.

Biggs, R., Schlüter, M., & Schoon, M. L. (2015). An introduction to the resilience approach and principles to sustain ecosystem services in social-ecological systems. In R. Biggs, M. Schlüter, & M. L. Schoon (Eds.), *Principles for building resilience: Sustaining ecosystem services in social-ecological systems* (pp. 1–31). Cambridge, MA: Cambridge University Press. https://doi.org/10.1017/CBO9781316014240.002.

Bulkeley, H., & Castán Broto, V. (2013). Government by experiment? Global cities and the governing of climate change. *Transactions of the Institute of British Geographers, 38*(3), 361–375. https://doi.org/10.1111/j.1475-5661.2012.00535.x.

Burrough, P. A., Wilson, J. P., Van Gaans, P. F., & Hansen, A. J. (2001). Fuzzy k-means classification of topo-climatic data as an aid to forest mapping in the Greater Yellowstone Area, USA. *Landscape Ecology, 16*(6), 523–546. https://doi.org/10.1023/A:1013167712622.

Dakos, V., Quinlan, A., Baggio, J. A., Bennett, E., Bodin, Ö., & Burnsilver, S. (2015). Principle 2 – manage connectivity. In R. Biggs, M. Schlüter, & M. L. Schoon (Eds.), *Principles for building resilience: Sustaining ecosystem services in social-ecological systems* (pp. 80–104). Cambridge, MA: Cambridge University Press. https://doi.org/10.1017/CBO9781316014240.002.

Deser, C., Phillips, A., Bourdette, V., & Teng, H. (2012). Uncertainty in climate change projections: The role of internal variability. *Climate Dynamics, 38*(3–4), 527–546. https://doi.org/10.1007/s00382-010-0977-x.

FOCC (Florida Oceans and Coastal Council). (2010). *Climate change and sea-level rise in Florida: An update of the effects of climate change on Florida's ocean and coastal resources.* Tallahassee, FL: FOCC (Report).

Folke, C. (2006). Resilience: The emergence of a perspective for social-ecological systems analyses. *Global Environmental Change, 16*(3), 253–267. https://doi.org/10.1016/j.gloenvcha.2006.04.002.

Folke, C., Hahn, T., Olsson, P., & Norberg, J. (2005). Adaptive governance of socio-ecological systems. *Annual Review of Environment and Resources, 30*(1), 441–473. https://doi.org/10.1146/annurev.energy.30.050504.144511.

Gorsevski, P. V., Gessler, P. E., & Jankowski, P. (2003). Integrating a fuzzy k-means classification and a Bayesian approach for spatial prediction of landslide hazard. *Journal of Geographical Systems, 5*(3), 223–251. https://doi.org/10.1007/978-3-642-03647-7_31.

Gray, V. (2007). Climate change 2007: The physical science basis summary for policymakers. *Energy and Environment, 18*(3–4), 433–440. https://doi.org/10.1260/095830507781076194.

Hallegatte, S. (2009). Strategies to adapt to an uncertain climate change. *Global Environmental Change, 19*(2), 240–247. https://doi.org/10.1016/j.gloenvcha.2008.12.003.

Handmer, J. W., & Dovers, S. R. (1996). A typology of resilience: Rethinking institutions for sustainable development. *Industrial and Environmental Crisis Quarterly, 9*(4), 482–511. https://doi.org/10.1177/108602669600900403.

Hawken, P. (2017). *Drawdown: The most comprehensive plan ever proposed to reverse global warming.* New York, NY: Penguin Books.

He, Z. (2006). *Farthest-point heuristic based initialization methods for k-modes clustering*. CoRR, abs/cs/0610043.

Hilborn, R. (1987). Living with uncertainty in resource management. *North American Journal of Fisheries Management, 7*, 1–5. https://doi.org/10.1577/1548-8659(1987)7<1:LWUIRM>2.0.CO;2.

Holling, C. S. (1973). Resilience and stability of ecological systems. *Annual Review of Ecology and Systematics, 4*(1), 1–23. https://doi.org/10.1146/annurev.es.04.110173.000245.

Holling, C. S., Folke, C., Carpenter, S., Walker, B., Scheffer, M., Elmqvist, T., & Gunderson, L. (2004). Regime shifts, resilience, and biodiversity in ecosystem management. *Annual Review of Ecology, Evolution, and Systematics, 35*, 557–581. https://doi.org/10.1146/annurev.ecolsys.35.021103.105711.

Homer-Dixon, T., Walker, B., Biggs, R., Crépin, A.-S., Folke, C., Lambin, E. F., … Troell, M. (2015). Synchronous failure: The emerging causal architecture of global crisis. *Ecology and Society, 20*(3), 6. https://doi.org/10.5751/ES-07681-200306.

Huang, Z. (1997). A fast clustering algorithm to cluster very large categorical data sets in data mining. In A. Z. Tucson (Ed.), *Paper presented at the workshop on research issues on data mining and knowledge discovery*. Retrieved from http://citeseerx.ist.psu.edu/viewdoc/summary?doi=10.1.1.6.4718.

Ji, J., Pang, W., Zheng, Y., Wang, Z., Ma, Z., & Zhang, L. (2015). A novel cluster center initialization method for the k-prototypes algorithms using centrality and distance. *Applied Mathematics and Information Sciences, 9*(6), 2933–2942. https://doi.org/10.12785/amis/090621.

Jones, R. N. (2000). Managing uncertainty in climate change projections: Issues for impact assessment. *Climatic Change, 45*(3), 403–419. https://doi.org/10.1023/A:1005551626280.

Ketchen, D. J., & Shook, C. L. (1996). The application of cluster analysis in strategic management research: An analysis and critique. *Strategic Management Journal, 17*(6), 441–458. https://doi.org/10.1002/(SICI)1097-0266(199606)17:6<441::AID-SMJ819>3.0.CO;2-G.

Kotschy, K., Biggs, R., Daw, T., Folke, C., & West, P. (2015). Principle 1: Maintain diversity and redundancy. In R. Biggs, M. Schlüter, & M. Schoon (Eds.), *Principles for building resilience: Sustaining ecosystem services in social-ecological systems* (pp. 50–79). Cambridge: Cambridge University Press. https://doi.org/10.1017/CBO9781316014240.004.

Latif, M. (2011). Uncertainty in climate change projections. *Journal of Geochemical Exploration, 110*(1), 1–7. https://doi.org/10.1016/j.gexplo.2010.09.011.

Lee, K. (1999). Appraising adaptive management. *Conservation Ecology, 3*(2). https://doi.org/10.5751/ES-00131-030203.

Li, Y., Shi, Z., Li, F., & Li, H. (2007). Delineation of site-specific management zones using fuzzy clustering analysis in a coastal saline land. *Computers and Electronics in Agriculture, 56*(2), 174–186. https://doi.org/10.1016/j.compag.2007.01.013.

MacQueen, J. (1967). Some methods for classification and analysis of multivariate observations. In L. M. Le Cam, & J. Neyman (Eds.), *Proceedings of the 5th Berkeley Symposium on Mathematical Statistics and Probability* (pp. 281–297). Berkley, CA: University of California Press.

Margolis, S. E., & Liebowitz, S. J. (1995). Path dependence, lock-in, and history. *Journal of Law, Economics, and Organization, 11*(1), 205–226. https://doi.org/10.1093/oxfordjournals.jleo.a036867.

Matin, N., Forrester, J., & Ensor, J. (2018). What is equitable resilience? *World Development, 109*, 197–205. https://doi.org/10.1016/j.worlddev.2018.04.020.

Maulik, U., & Bandyopadhyay, S. (2002). Performance evaluation of some clustering algorithms and validity indices. *IEEE Transactions on Pattern Analysis and Machine Intelligence, 24*(12), 1650–1654. https://doi.org/10.1109/TPAMI.2002.1114856.

McGinnis, M. D., & Ostrom, E. (2014). Social-ecological system framework: Initial changes and continuing challenges. *Ecology and Society, 19*(2). https://doi.org/10.5751/ES-06387-190230.

Moser, S. C. (2005). Impact assessments and policy responses to sea-level rise in three US states: An exploration of human-dimension uncertainties. *Global Environmental Change, 15*(4), 353–369. https://doi.org/10.1016/j.gloenvcha.2005.08.002.

Nelson, D. R. (2011). Adaptation and resilience: Responding to a changing climate. *Wiley Interdisciplinary Reviews: Climate Change, 2*(1), 113–120. https://doi.org/10.1002/wcc.91.

Nelson, D. R., Adger, W. N., & Brown, K. (2007). Adaptation to environmental change: Contributions of a resilience framework. *Annual Review of Environment and Resources, 32*(1), 395–419. https://doi.org/10.1146/annurev.energy.32.051807.090348.

Netzel, P., & Stepinski, T. (2016). On using a clustering approach for global climate classification. *Journal of Climate, 29*(9), 3387–3401. https://doi.org/10.1175/JCLI-D-15-0640.1.

Pachauri, R. K., & Meyer, L. A. (Eds.). (2014). *Climate Change 2014: Synthesis Report.* Geneva: IPCC. Contribution of Working Groups I, II and III to the Fifth Assessment Report of the Intergovernmental Panel on Climate Change.

Peterson, G., De Leo, G. A., Hellmann, J. J., Janssen, M. A., Kinzig, A., Malcolm, J. R., ... Tinch, R. R. T. (1997). Uncertainty, climate change, and adaptive management. *Conservation Ecology, 1*(2).

Pike, A., Dawley, S., & Tomaney, J. (2010). Resilience, adaptation and adaptability. *Cambridge Journal of Regions, Economy and Society, 3*(1), 59–70. https://doi.org/10.1093/cjres/rsq001.

Rahmstorf, S. (2007). A semi-empirical approach to projecting future sea-level rise. *Science, 315*(5810), 368–370. https://doi.org/10.1126/science.1135456.

Scheffer, M., Carpenter, S. R., Lenton, T. M., Bascompte, J., Brock, W., Dakos, V., ... Vandermeer, J. (2012). Anticipating critical transitions. *Science, 338*(6105), 344. https://doi.org/10.1126/science.1225244.

Schoon, M., Robards, M., Brown, K., Engle, N., Meek, C., & Biggs, R. (2015). Politics and the resilience of ecosystem services. In R. Biggs, M. Schlüter, & M. Schoon (Eds.), *Principles for building resilience: Sustaining ecosystem services in social-ecological systems* (pp. 32–49). Cambridge: Cambridge University Press. https://doi.org/10.1017/CBO9781316014240.003.

Smith, J., & Saito, M. (2001). Creating land-use scenarios by cluster analysis for regional land-use and transportation sketch planning. *Journal of Transportation and Statistics, 4*(1), 39–47.

Sorensen, A., & Hess, P. (2015). Building suburbs, Toronto-style: Land development regimes, institutions, critical junctures and path dependence. *Town Planning Review, 86*(4), 411–436. https://doi.org/10.3828/tpr.2015.26.

Stankey, G. H., Clark, R. N., & Bormann, B. T. (2005). *Adaptive management of natural resources: Theory, concepts, and management institutions.* General Technical Report PNW-GTR-654. Portland, OR: United States Department of Agriculture. https://doi.org/10.2737/PNW-GTR-654.

Steffen, W., Richardson, K., Rockström, J., Cornell, S. E., Fetzer, I., Bennett, E. M., ... Sörlin, S. (2015). Planetary boundaries: Guiding human development on a changing planet. *Science, 347*(6223). https://doi.org/10.1126/science.1259855.

Walker, B., Kinzig, A., & Langridge, J. (1999). Plant attribute diversity, resilience, and ecosystem function: The nature and significance of dominant and minor species. *Ecosystems, 2*(2), 95–113. https://doi.org/10.1007/s100219900062.

Walker, B., & Salt, D. (2012). *Resilience practice: Building capacity to absorb disturbance and maintain function.* Washington, DC: Island Press.

Xu, H., Ma, C., Lian, J., Xu, K., & Chaima, E. (2018). Urban flooding risk assessment based on an integrated k-means cluster algorithm and improved entropy weight method in the region of Haikou, China. *Journal of Hydrology, 563,* 975–986. https://doi.org/10.1016/j.jhydrol.2018.06.060.

Zimmerman, R., & Faris, C. (2011). Climate change mitigation and adaptation in North American cities. *Current Opinion in Environmental Sustainability, 3*(3), 181–187. https://doi.org/10.1016/j.cosust.2010.12.004.

The links between vulnerability, poverty, and natural hazards: A focus on the impacts of globalization trends

Mahmood Fayazi[1], Lisa Bornstein[2]

[1]The Institute for Disaster Management and Reconstruction, Sichuan University and the Hong Kong Polytechnic University, Chengdu, Sichuan, China; [2]School of Urban Planning, McGill University, Montreal, Quebec, Canada

OUTLINE

1. Introduction

We live in a world characterized by increasing frequency, severity, and intensity of hazardous events that make communities suffer physically, economically, and emotionally. Hazard-related studies argue that the likelihood of disaster losses is based on the interplay between exposure to an external threat (such as a natural hazard) and internal conditions broadly called vulnerabilities (Holloway, 2006). Although it is commonly accepted that vulnerability is "the characteristics and circumstances of a community, system or asset that make it susceptible to the damaging effects of an event" (UNISDR, 2013), the origins of vulnerability are often a matter of debate. Hazard-related studies persistently consider vulnerability as a failure to access critical soft and hard resources, leading communities to unsafe conditions and putting people and assets at risk (Blaikie, Cannon, Davis, & Wisner, 1994; Lizarralde, Johnson, & Davidson, 2010). In contrast, development researchers argue that repeating shocks caused by natural hazards reduce the economic power of households, communities, and countries, creating poverty and exacerbating the vulnerability to future disasters (Pelling, 2003). These observations suggest that it is theoretically possible to specify a cycle of vulnerability in which poverty contributes to vulnerability to disasters, and again the resulted losses exacerbate poverty and other social ills. This chapter reviews different understandings of vulnerability—to natural hazards and to poverty—and proposes a framework that can be employed in natural-hazard and development studies. Such an analytical framework combines the concepts of vulnerability and poverty extracted from natural-hazard and development studies. In this framework, two sources of disturbances affect vulnerable communities. On one side, the adverse impacts of globalization trends, taken as examples of disturbing external events, drive the vulnerable to the state of poverty, and on the other side, natural hazards push the poor and exacerbate their vulnerabilities to disasters. This is a cyclic condition moving from vulnerability to poverty, to vulnerability to natural hazards repeatedly.

This chapter is divided into six sections. First, the relevant literature on vulnerability to hazards and poverty is reviewed: the "pressure and release model" (Blaikie et al., 1994), the entitlement theory (Sen, 1997), and a recent revision of the concept of vulnerability (Miller et al., 2010). The second section presents a theoretical model, explaining the cycle of vulnerability creation from vulnerability to poverty to vulnerability to natural hazards and reverse. The third section describes the research methods. Then, some cases are explored to show how global changes in general—and globalization trends in particular—can lead communities to a state of poverty. Fifth, cases drawn from the literature are used to illustrate how natural hazards exacerbate communities' poverty and increase their vulnerabilities to future events. Finally, in the section of discussion and conclusion, the principal findings and their theoretical and practical implications are presented.

1.1 Poverty

Economists tend to assess poverty according to absolute or relative measures of access to material resources, which define, for instance, a poverty line (Adger, 1997; Baulch, 1996; Blackwood & Lynch, 1994). Adger (1999), however, argues that resources and wealth are mediated through property rights and access to them, which are also conditional on other determinants such as social and economic relations. In fact, social studies recognize multifaceted forms of poverty and explain their social, political, and economic causal factors. For instance, Beall (2002, p. 44) claims that "the conditions of life depriving people from social participation and the exercise of full citizenship" impact the abilities to use resources required to secure livelihood.

Development literature and studies on poverty, famine, and food insecurity define vulnerability as a proneness to a sudden fall in the level of welfare or poor access to enough food. It is often argued that poverty creates vulnerability and is a result of it (Ellis, 2006). On the one hand, poverty is a critical factor that pushes communities closer to the edge of vulnerability to famine, food insecurity, and natural hazards. On the other hand, poverty is an adverse outcome of vulnerability to undesirable events that decrease access to commodities and resources. However, the poor may not be necessarily vulnerable if they live in a relatively safe context with a strong, stable, social support system, as Ellis (2006), among others, recognizes. Likewise, nonpoor people might be extremely vulnerable if exposed to threats to which they cannot easily adapt and for which their assets are ill-equipped to handle. Therefore, the relationship between poverty and vulnerability to poverty still remains controversial, and any answer to their causal factors depends on a profound understanding of dynamic contextual conditions in which various factors must be taken into consideration.

1.2 Vulnerability

Different disciplines, from economics and anthropology to psychology and engineering, use the concept of vulnerability, albeit in different ways. Studies on environmental resources and risks adopt it to describe socioecological systems, as well as vulnerabilities to poverty and deprivation (Adger, Paavola, Huq, & Mace, 2006). These two main research traditions, which examine the use of environmental resources and environmental risks, consider two dimensions of vulnerability: to natural hazards and to poverty.

In terms of vulnerability to natural hazards, Burton et al. consider the exposure to, the probability of, and the intensity of damaging events to explain the different levels of vulnerability that various groups in society may have (Burton, Kates, & White, 1978). These variables concentrate on one aspect of the vulnerability equation, namely the attributes of the natural event. Later contributions in the field of human ecology, however, critique this approach to vulnerability focused on these attributes alone. They argue that this approach fails to consider the political, cultural, and structural conditions of the society, which, in their view, are the source of significant vulnerabilities. Defenders of the second approach find in these social attributes the real causes of the vulnerability of the poor and marginalized (Hewitt, 1997; Watts, 1983). According to this approach, poor households are not vulnerable because they live in hazardous areas, but because social, economic, and political pressures push them to occupy these unsafe locations and expose them to natural events (Hewitt, 1994, 1997).

The "pressure and release model" developed by Blaikie et al. explains why and how unsafe conditions are created (Blaikie et al., 1994). These authors explain that root causes (often historic, economic, political, and social conditions) lead societies to dynamic pressures such as rapid rural migration, lack of infrastructure, and poverty. These pressures eventually materialize in unsafe conditions that put people and assets at risk (Blaikie et al., 1994). The model provided a valuable contribution, demonstrating that disasters are not really "natural" but caused by human action. Despite all its merits, it has received some criticism. Adger (2006) claims that the "pressure and release model" provides a mechanistic view of vulnerability.

According to him "operationalizing the pressure and release model [...] involves typologies of causes [...], limiting the analysis in terms of quantifiable or predictive relationships" (Adger, 2006). Similarly, Lizarralde et al. (2010, p. 3) write that "this model indicates very little about what type of actions are required to overcome the disaster once the natural event coincides with the accumulated vulnerability."

These authors define the concept of vulnerability as a lack of access to two types of resources: hard resources (such as housing, roads, infrastructure, and public services) and soft resources (such as decision-making capacity, employment, education, and information) (Lizarralde et al., 2010). Lizarralde et al. argue that vulnerability corresponds to inappropriate or insufficient access to the resources that allow a community to deal with the effects of hazards. Then, based on this conception of vulnerability, they define post-disaster reconstruction projects as a process of "improving access to resources that have been lost and developing access to the basic resources that people probably did not even have before the disaster" (Lizarralde et al., 2010, p. 10). This conception of vulnerability, however, indicates little about the origins or the causes of insufficient access to resources. They also hardly distinguish between the lack of access to resources that create vulnerabilities (before the disaster) and that which results from the disaster.

Sen's "entitlement theory" provides some answers (Sen, 1982, 1997). It addresses the relationships between access to resources and vulnerability, examining, among other dynamics, the insufficient and inappropriate access to rights and resources that can result from undesirable events. Entitlement theory considers the vulnerability of households to poverty as a set of linked economic and institutional factors (Leach, Mearns, & Scoones, 1997). For Sen, entitlements are the set of alternative commodity bundles that a person can command in a society using the totality of rights and opportunities that he or she has (Sen, 1997).

Examining these studies together allows one to theoretically imagine a cycle of vulnerability in which poverty contributes to vulnerability to natural hazards, which then, when a "disaster" occurs, generates adverse outcomes such as inequalities, exclusion, and lack of access to rights and resources. These adverse outcomes further exacerbate vulnerabilities and lead communities into increased poverty communities were originally poor, so poverty is increased rather than instigated. Given the importance of understanding the mechanisms of disaster risk reduction (Fayazi et al., 2017; UNISDR, 2005), the cross-fertilization of the entitlement theory with the concept of vulnerability seems likely to yield important insights.

2. Theoretical model

On the one hand, poverty can transform a natural event into a disaster, as experienced by specific individuals, households, or communities. On the other hand, disasters undoubtedly affect communities' access to resources, exacerbating or creating poverty. Besides, post-disaster interventions under recovery and reconstruction programs often cause inequalities, benefiting some groups of households while leaving others more vulnerable (Fayazi & Lizarralde, 2017). Less-vulnerable families often have more access to resources allocated to reconstruction and recovery programs, while the poor are sometimes excluded. As such, pre-disaster vulnerable communities not only suffer from natural hazards but also experience considerable difficulties in reconstruction and recovery programs. They are also often excluded from decision-making processes and may have societally constructed restrictions on their access to certain benefits. Unsurprisingly, these deprivations make excluded and unentitled communities vulnerable to large- or even small-scale undesirable events. For instance, global or national economic policies as examples of adverse pressures from outside of communities can easily harm small businesses in vulnerable communities and eventually push them to a state of poverty. These poor households seek to escape poverty through, for example, migration to informal settlements that are often characterized by unsafe dwellings. This cycle of vulnerability creation, exacerbation, and transformation is presented in the model of "vulnerability creation and exacerbation" (Fig. 14.1).

The diagram borrows the basic principles of the "pressure and release model" (Blaikie et al., 1994), the "entitlement theory" (Sen, 1997), and the revision of the concept of vulnerability proposed by Lizarralde et al. (2010). The model seeks to illustrate how social exclusion and lack of entitlement increase the vulnerability of communities to poverty, and how poverty leads communities to dynamic pressures and eventually materializes in unsafe conditions where people and assets are at risk of natural hazards. In other words, this model links *vulnerability to poverty* with *vulnerability to natural hazards*. Indeed, this model yields significant insight as to how the vulnerability is generated, exacerbated, and turned into another form of more extreme vulnerability.

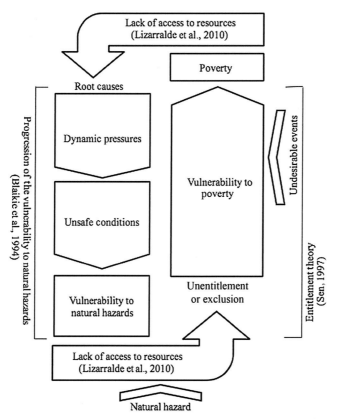

FIGURE 14.1 Vulnerability creation and exacerbation model. *Credit: The Authors.*

3. Research method

This study was conducted as a final assignment for the course "Cities in a Globalizing World" in the School of Urban Planning at McGill University in Montreal, Canada. The report for the final assignment was then developed into this book chapter. This research draws primarily on the review of relevant literature in the hazard and development-related fields. Based on an extensive review of the literature, this study develops a theoretical model, and to validate it, this chapter draws on a number of cases detailed in the literature. In the preliminary steps, 56 publications from different fields of study were selected. Qualitative content analysis and a categorization technique were used to draw up a shortlist of ten case studies. After an in-depth review of these cases, four were then chosen from the shortlist of case studies. In the beginning, the objective was to select case studies that support the complete circle of the vulnerability creation and exacerbation in the theoretical model presented in this study. However, except the case of Xuan Thuy district in Vietnam, which has experienced both globalization pressures and devastating typhoons, the others only support one side of the model. We then decide to select cases that present each side of the circle individually. First, we focus on globalization as an undesirable process that affects communities and creates exclusion, lack of entitlement, and poverty. The cases from Faisalabad in Pakistan and Barasat Town in West Bengal were chosen to explain the potential adverse impacts of globalization in vulnerable communities. Then, the case of the 1999 earthquakes in Marmara, Turkey, is used to support our argument and elucidate how poverty plays a critical role in exacerbating vulnerability to natural hazards. At the end, the case of Vietnam is presented to describe the vicious circle of vulnerability creation, transformation, and exacerbation.

4. From exclusion and unentitlement to poverty

Development studies refer to the lack of entitlement and social exclusion as the main sources of vulnerability to poverty. Doubtless, the communities with limited access to rights and resources find it difficult to maintain livelihoods, diversify their sources of income, and benefit from recovery mechanisms to mitigate the adverse effects of undesirable events (Fayazi & Lizarralde, 2018; Sen, 1997). These researchers believe that the principal components of vulnerability are the lack of entitlement and social exclusion. Given the fact that globalization trends are often tied to "undesirable events" (Beall, 2002; Dollar, 2001; Jenkins, 2004), this chapter starts by exploring the adverse impact of globalization on the marginalized and socially excluded communities to explain the process of poverty exacerbation.

4.1 Globalization

In recent decades, high-tech communication and information technologies have reduced distances and revealed a new scale at which information, capital, and knowledge can disseminate. High-speed transportation facilities further the transfer of goods, fuels, and labor. In this novel world, sources of wealth are argued to flow far more freely among and within countries, cities, and international firms than in the past (Jenkins, 2004). This trend creates, in fact, a network of capital, which is highly centralized and controlled by large corporations and, simultaneously, significantly decentralized (Castells, 1999). Small- and medium-size businesses find it difficult to compete with large-size and global ones. Booming profitable companies hire and join others to minimize competitors, win global competition, and remain in the network. Paradoxically, in this network, internationalized businesses find it difficult to succeed without the small ones. Businesses operate often not as single global corporations but rather as decentralized networks "whose elements are given considerable autonomy" (Castells, 1999, p. 6). These elements are often small scale and local businesses that are connected to several large corporations (Taylor, Walker, Catalano, & Hoyler, 2002). The "earned benefits" of the global capitalism, therefore, are not concentrated but are distributed among an increasing number of shareholders. However, values and interests—for products and services—change relentlessly (Jenkins & Wilkinson, 2002; Sassen, 2004). This creates an endless competition in which the winners are inside, and the losers are outside the network (Castells, 1999, p. 6).

This global competition leads to complex and unpredictable movements (of capital, technology, knowledge, and labor) that redefine the traditional role of nation states (Jenkins & Wilkinson, 2002). Some national or state governments band together, creating or adapting "supranational" institutions to manage the global complex and unpredictable changes. Others are forced to change their operations—the ways in which they relate to their policy and to other states—to remain in the global network, otherwise, they would be totally excluded. Likewise, this dynamic global network allows the systems to link up and disconnect valuable and devalued things. It has, in fact, a "capacity to include and exclude people, territories, and activities that characterize the new global economy" (Castells, 1999, p. 5). This exclusion raises additional challenges for those who are dependent on global networks and triggers social difficulties such as inequality, unemployment, crime, and poverty.

According to Castells (1999), the role of current globalization trends in exacerbating poverty includes four dimensions: individualization of labor, overexploitation, social exclusion, and perverse integration. Individualization is an inherent characteristic of globalization in which labor individually competes to contribute to production (Jenkins & Wilkinson, 2002). Castells (1999) warns of the danger of exclusion of labor force (especially those who are non and less skilled) from job opportunities because it eventually exacerbates social inequalities and poverty. Unfavorable labor conditions, such as poor wages and overexploitation, are imposed on fragile categories of workers such as immigrants and minorities. Social exclusion is associated with the globalization trend that bars individuals or groups from access to social benefits and resources (Castells, 1999). Individuals or groups who lose their position due to the global economic competition may find themselves in a state of social exclusion in which they have limited access to jobs and social welfare (Jenkins & Wilkinson, 2002). Finally, Castells points out that "numbers of people are being excluded from access to regular jobs, they are moving onto [a] floor of crime," a process that he calls perverse integration (Castells, 1999, p. 9).

Globalization trends, thus, impact individuals, communities, firms, and nations and can contribute to a state of poverty. To continue, this chapter discusses two cases from Faisalabad in Pakistan and Barasat Town in West-Bengal that explore how globalization trends affect communities and push them to a state of exacerbated social exclusion, unentitlement, and poverty.

4.2 Economic exclusion on the basis of social exclusion: The case of Faisalabad

Beall (2002) describes a process of economic deprivations arising from social ones in an article entitled "Globalization and social exclusion in cities: Framing the debate with lessons from Africa and Asia." Her research focuses on Faisalabad, Pakistan, and explores how globalization trends lead the socially excluded community of "sweepers" to a new form of economic exclusion. Faisalabad is a city of around two million people in the Punjab Province that is linked to global processes in the hope of reversing a deteriorating position in the international textile market. Recent globalization trends forced national and local governments in Pakistan to downsize the large municipal workforce devoted to "the cozy world" of municipal waste collection. The local government—municipality—had to downsize its structure to be able to adapt to changing values and interests in the global textile market and to offer essential support services. The most stigmatized and marginalized residents of Faisalabad were sweepers, who then lost their jobs and faced poverty.

The community of sweepers can be traced historically to the rural Hindu caste called Churha, customarily associated with "polluting" work. In Pakistan, certain types of work are still associated with tribal origins and "as a result, Punjabi Christian sweepers are regarded as polluted or unclean and, on this basis, work associated with drains, sewers and waste collection has remained their exclusive preserve, almost to this day" (Beall, 2002, p. 46).

This identity-based socially excluded community was economically integrated and had exclusive access to secure employment in the public sector, with jobs customarily passed from one generation to the next, involving multiple family members, and rarely passing outside the group. By using the prejudices associated with their caste origins, the sweepers had achieved and retained for themselves access to a secure form of livelihood.

However, global competition forced the municipality to downsize its workforce, impacting the access of sweepers to safe livelihood and employment. In the context of few employment opportunities available to the identity-based excluded community of sweepers, "there is evidence of increasing deprivation and insecurity among this group" (Beall, 2002, p. 47). The experience of the Punjabi sweepers illustrates how an undesirable event can push a socially excluded community (but one that had access to steady employment) to a state of economic exclusion and poverty.

4.3 Migration to urban slums due to globalization pressures: The case of Barasat Town

The second case draws on an article by Notan Bhusan Kar (2012), entitled "Impact of globalization on urban slum dwellers-migrants from rural areas: A social case study of Barasat Town, North 24-Parganas District, West Bengal." In 1990s, the Indian national government, in line with World Bank and International Monetary Fund recommendations, adopted economic policies toward globalization, liberalization, and privatization. These organizations pushed the government to withdraw subsidies from the agricultural sector. As a result,

Kar argues, the cost of agricultural production increased, making traditional farming an unprofitable option for Indian farmers. The increase in the cost of the production sector forced thousands of farmers to give up their agricultural occupations. In addition, land acquisitions promoted by developmental projects rendered many peasants landless and jobless. A large number of now poor and marginal peasants were forced to opt for other means of livelihood, and many migrated to urban areas looking for subsistence jobs. Most of them were unskilled, thus jobs—where obtained—were paid poorly, and housing costs in urban areas forced them to live in slums.

This case shows how globalization pressures can exclude local communities from benefits and push them to a state of poverty. These examples support the first part of our theoretical model (the right side of the model) that illustrates how an undesirable event, under the influence of globalization trends, threatens a community and can lead to exclusion, unentitlement, and poverty. The discussion now turns to the second part of the model.

5. From poverty to natural hazards

In the international academic arena, there is an increasing awareness of the integral link between the reduction of poverty and that of the risk of natural hazards (Norris, Stevens, Pfefferbaum, Wyche, & Pfefferbaum, 2008). Extreme poverty, according to UN-Habitat (2003), is the main source of vulnerability to damaging natural events. Poverty is an important aspect of vulnerability to natural hazards as it directly influences vulnerability indicators: exposure, sensitivity, and coping capacity (Burton et al., 1978; Gallopin, 2003). In developing countries, extremely poor families and individuals have only access to low-cost housing in squatter settlements located in hazardous areas. Marginalization, as another type of deprivation, is fundamentally associated with poverty and impacts emergency coping capacities. UN-Habitat (2003) argues that slums are often not recognized by public authorities as an integral or equal part of the city. They often lack essential services that cause significant vulnerabilities to disease and accidents (Mutisya & Yarime, 2011). According to Adger (1999), poverty also hinders recovery from the impacts of hazards. Poor people are often the ones to suffer more from hazards but also those who have less capacity to recover (Kahn, 2005).

The global assessment of slums undertaken by UN-Habitat (2003) shows that squatter settlements are often found in high-risk areas, such as steep hill slopes, deep gullies, near dumpsites, and in flood-prone areas. These findings are consistent with what Lizarralde (2015) observed after the investigation of slums in Asia, Latin America, Central America, and Africa. Similarly, Holloway (2006) explains how the rapid growth of cities in developing countries pushes the poor and the socially excluded community to occupy and create squatter settlements, which eventually increases their vulnerability to natural hazards. UN-Habitat (2007, p. 27) explains the concept of cumulative vulnerabilities: the poor who live in hazardous locations are forced to accept risks as they cannot afford safe places in the city. To avoid the risk of eviction from formal areas, they have to take other risks. Research on Marmara, Turkey, illustrates how poor and excluded communities are vulnerable to, and suffer from, natural hazards, providing support for the second part of our theoretical model.

5.1 The role of the rapid growth of cities in the creation of poverty and exacerbation of vulnerability to natural hazards: The 1999 earthquakes in Marmara, Turkey

Material from the piece by Johnson (2007b) on "Strategic planning for postdisaster temporary housing" provides an indication of the cycle of linkages among poverty, vulnerability, "hazardous" events, and compounded hardship. According to Johnson (2007b), economic liberalization, industrial development, and political conflicts in south-east Turkey provoked rapid migration to cities. The mass migration of poor rural families led to uncontrolled urbanization and the creation of large-scale informal settlements; there was little compliance with building codes and little use of safe building techniques. Shortcuts in construction techniques eventually increased the vulnerability of inhabitants to natural hazards. As a result, two earthquakes in August and November 1999 (with 7.4 and 7.2 magnitudes in Richter scale) killed more than 18,000 individuals and destroyed over 300,000 houses.

After the earthquakes, the reconstruction programs exacerbated social gaps and generated new vulnerabilities. Three to six months following the earthquake, the Turkish Government, in collaboration with private organizations, provided about 42,000 prefabricated units to serve as transitional shelters before the reconstruction of permanent houses. The units were mostly placed on the outskirts of urban areas, requiring the construction of roads, community centers, bus services, postal services, and commerce. The 30-m^2 prefabricated duplex units were constructed on slab foundations with corrugated iron roofs and included a kitchen and a bathroom with running water (Johnson, 2007a). Although the transitional sheltering program provided safe and comfortable shelters for people in need and helped them to return to their normal life, the approach adopted for locating the units outside city limits led to urban sprawl and the emergence of new informal settlements. As Johnson, Lizarralde, and Davidson (2006) observed five years after the earthquake, most housing units were still standing, and many of them occupied as rental housing. To avoid political trouble and public backlash, the authorities did not pull the transitional shelters down and restrained from evicting the occupants. On the other hand, rental prices for an apartment unit raised more than double, as the earthquake and transitional shelters provided lodging for families who were squeezed out of the rental market. Consequently, the reconstruction program widened social gaps, marginalized the poor, and rendered new forms of vulnerability to both poverty and natural hazards.

Lizarralde et al. (2010) make similar observations about the creation of vulnerability in Turkey. They argue that the mass migration of low-income rural people to cities and the expansion of informal settlements became dynamic pressures that had root-causes in adverse industrial and economic policies that created extreme vulnerabilities. In sum, the processes of poverty creation generate dynamic pressures and generate vulnerability to hazards.

5.2 Inequality, poverty, and social vulnerability to climate change in coastal Vietnam

Adger (1999) explains how social and economic factors associated with an economic transition from central planning led to the increased inequalities and social vulnerabilities to climate change and climate extremes in coastal communities in Northern Vietnam. This case covers the two sides of our framework. This study explains the impact of globalization pressure in rendering inequalities (right side of the vulnerability creation and exacerbation model) and discriminating the marginalized population by restricting access to recovery resources after the periodical landfall typhoons (left side of the model).

Xuan Thuy district is a relatively prosperous area of northern Vietnam, with a small proportion of poor people who fall well behind the average economic class of the population in the country. The relative prosperity indicates access to fertile agricultural land and productive coastal resources, while the extent of poverty coincides with lower land allocation and reliance on commonly managed resources. Adger (1999) explores the fact that the sources of income for the poor and nonpoor sections of Xuan Thuy's population are significantly different. The poor rely on salt-making in the coastal communities of the south where there is a significant agricultural land scarcity. However, the nonpoor households rely to a greater extent on rice and agriculture for their main source of income, and they are also engaged in commercial aquaculture. The income inequality serves as a proxy for understanding inequality of access to resources, which is directly correlated to a concentration of the ownership of productive assets in a small proportion of the population.

According to Adger (1999), pressures of globalization resulted in market liberalization and legalization of private enterprises, which consequently increased inequalities in northern Vietnam. Despite interventions in agricultural inputs and markets as well as the planned allocation of labors during the Đổi Mới economic reforms, Vietnam has experienced interregional increases in inequality. Besides, social hierarchies and distribution of formal political power within communities played crucial roles in exacerbating inequalities. All in all, Adger (1999) reveals shreds of evidence of the fact that individuals at the lower end of the distribution of income are discriminated regarding access to recovery resources in disaster situations. Xuan Thuy district often encounters the most significant number of typhoons and storm surges on average, in September and October, when the monsoonal current changes direction from southwest to northeast (Kelly & Adger, 2000). In this area, the community and its economic assets such as agriculture and aquaculture experience extensive damage caused by storm surges, sea-level rise, as well as high waves, and strong winds every couple of years. Since the beginning of the 20th century, as a direct result of extended floods, typhoons, and epidemics, nearly 25,000 lives have been lost in Vietnam (Kleinen, 2007).

The storms pose enormous risks for communities, and the poor are more susceptible to the impacts of landfall typhoons. The reduction in power and autonomy of state institutions to manage collective measures for protection from the impacts of coastal storms is a significant channel of vulnerability in Xuan Thuy. Due to the privatization of enterprises, the state institutions are no longer capable of allocating labor and resources to manage water and irrigation for coastal defense. The privatization has, however, contributed to increase marketed production of agricultural commodities, thereby contributing to exacerbating inequality in the District. The rolling back of the state has been at some cost to collective security, more specifically for the poor because they have little diversity in their income source and fewer reserves to absorb shocks (Adger, 1999). The poor have reduced access to resources for coping with extreme events such as credit sources, and they are more reliant on activities such as salt-making, which could potentially experience a significant impact in the face of coastal flooding (Kleinen, 2007). According to Adger (1999), the poorer households tend to have fewer sources of income, in particular from aquaculture, wage labor, and remittances, and less land for rice production, which minimizes their coping capacity and increases their vulnerabilities to future storms.

6. Discussion and conclusion

This study proposes the framework of "vulnerability creation and exacerbation" to explain the relationship between the vulnerability of societies to economic and natural shocks. By doing so, it adopts vulnerability theories from development and hazard-related studies and examines their relationships using the concepts of "poverty" and "entitlement." The global trends unveil the vulnerability of socially excluded and unentitled communities to economic and political shocks, pushing them to poverty. The study suggests that the creation of economic vulnerability and its final outcomes have a direct impact on the creation of vulnerability to natural hazards.

The integration of development and natural-hazard studies reveals interlinked causes of vulnerabilities and yields insight into their relationships. This study explores the relentless cycle of vulnerability creation and exacerbation. The cycle might begin from vulnerability to poverty when a community is unentitled and socially excluded. The economic, political, and social pressures (such as global trends) put the system at extreme poverty, creating accumulated vulnerabilities to natural hazards. In fact, the pressures exacerbate poverty, increase social gaps, marginalize the poor, and prevent them from accessing critical resources to cope with and recover after natural hazards. Therefore, undesirable natural events adversely affect the poorest of the poor, who are often ill-prepared to deal with the events. For the poor, a hurricane, an earthquake, or a drought can mean a permanent submersion in poverty that increases vulnerabilities to other kinds of events.

The cases explored support the theoretical discussion and show how the processes of creation and exacerbation of vulnerability occur. This study confirms the integral links between different kinds of vulnerabilities. In fact, the circle of "vulnerability creation and exacerbation" shows how a community may experience undesirable outcomes of economic, social, political, and environmental events one after the other. Also, the "vulnerability creation and exacerbation" framework reveals the fact that by controlling causal factors of vulnerability to a kind of pressure, the system will be less vulnerable to other pressures. In other words, if a community increases its capacity to cope with economic, social, and political pressures, it will be more capable to withstand environmental shocks too. Overall, the findings help academics, practitioners, and decision-makers understand the origins of vulnerabilities and develop solutions to manage and reduce vulnerabilities and their possible adverse impacts.

Even though the proposed theoretical model is confirmed by the cases, additional studies are required to validate the framework and make it as precise and accurate as possible. This study, in fact, attempts to open a new avenue for reflection and debate, pushing forward the frontiers of knowledge on vulnerability and resilience. Scholars and practitioners still need to refine frameworks and units of assessment to understand the origins of disasters and develop strategies that facilitate the emergence of social justice.

Acknowledgments

We gratefully acknowledge the support from the Disaster Resilience and Sustainable Reconstruction Research Alliance (Observatoire Universitaire de la Vulnérabilité, la Résilience et la Reconstruction Durable) and the Fundamental Research Funds for the Central Universities in China.

References

Adger, W. N. (1997). *Income inequality in former centrally planned economies: Results for the agricultural sector in Vietnam* (*CSERGE GEC Working Paper*). Retrieved from http://citeseerx.ist.psu.edu/viewdoc/download?doi=10.1.1.663.7022&rep=rep1&type=pdf.

Adger, W. N. (1999). Social vulnerability to climate change and extremes in coastal Vietnam. *World Development, 27*(2), 249—269. https://doi.org/10.1016/S0305-750X(98)00136-3.

Adger, W. N. (2006). Vulnerability. *Global Environmental Change, 16*(3), 268—281. https://doi.org/10.1016/j.gloenvcha.2006.02.006.

Adger, W. N., Paavola, J., Huq, S., & Mace, M. J. (Eds.). (2006). *Fairness in adaptation to climate change.* Cambridge, MA: The MIT Press.

Baulch, B. (1996). Neglected trade-offs in poverty measurement. *IDS Bulletin, 27*(1), 36—42. https://doi.org/10.1111/j.1759-5436.1996.mp27001004.x.

Beall, J. (2002). Globalization and social exclusion in cities: Framing the debate with lessons from Africa and Asia. *Environment and Urbanization, 14*(1), 41—51. https://doi.org/10.1177/095624780201400104.

Blackwood, D., & Lynch, R. G. (1994). The measurement of inequality and poverty: A policy maker's guide to the literature. *World Development, 22*(4), 567—578. https://doi.org/10.1016/0305-750X(94)90112-0.

Blaikie, P. C., Cannon, T., Davis, I., & Wisner, B. (1994). At risk: Natural hazards, people's vulnerability, and disasters. London: Routledge.

Burton, I., Kates, R. W., & White, G. F. (1978). The environment as hazard. New York, NY: Oxford University Press.

Castells, M. (1999). *Information technology, globalization and social development (UNRISD discussion paper No. 114).* Geneva: United Nations Research Institute for Social Development.

Dollar, D. (2001). *Globalization, inequality, and poverty since 1980. Background paper.* Washington, DC: The World Bank

Ellis, F. (2006). Vulnerability and coping. In D. A. Clark (Ed.), *The Elgar companion to development studies* (pp. 671—675). Cheltenham: Edward Elgar.

Fayazi, M., Arefian, F. F., Gharaati, M., Johnson, C., Lizarralde, G., & Davidson, C. (2017). Managing institutional fragmentation and time compression in post-disaster reconstruction—the case of Bam. *International Journal of Disaster Risk Reduction, 21*, 340—349. https://doi.org/10.1016/j.ijdrr.2017.01.012.

Fayazi, M., & Lizarralde, G. (2017). *The impact of relocation policies on different household groups—the case of Bam.* Paper presented at the 8th i-Rec Conference 2017: Reconstruction and Recovery for Displaced Populations and Refugees, Toronto.

Fayazi, M., & Lizarralde, G. (2018). Conflicts between recovery objectives: The case of housing reconstruction after the 2003 earthquake in Bam, Iran. *International Journal of Disaster Risk Reduction, 27*, 317—328. https://doi.org/10.1016/j.ijdrr.2017.10.017.

Gallopin, G. (2003). A systemic synthesis of the relations between vulnerability, hazard, exposure and impact, aimed at policy identification. In ECLAC (Economic Commission for Latin America and the Caribbean) (Ed.), *Handbook for estimating the socio-economic and environmental effects of disasters* (pp. 2—5). Mexico, DF: ECLAC, LC/MEX/GS.

Hewitt, K. (1994). "When the great planes came and made ashes of our city…": Towards an oral geography of the disasters of war. *Antipode, 26*(1), 1—34. https://doi.org/10.1111/j.1467-8330.1994.tb00229.x.

Hewitt, K. (1997). Regions of risk: A geographical introduction to disasters. London: Longman.

Holloway, A. (2006). Disaster mitigation. In D. A. Clark (Ed.), *The Elgar companion to development studies* (pp. 130—135). Cheltenham: Edward Elgar.

Jenkins, P., & Wilkinson, P. (2002). Assessing the growing impact of the global economy on urban development in Southern African cities: Case studies in Maputo and Cape Town. *Cities, 19*(1), 33—47. https://doi.org/10.1016/S0264-2751(01)00044-0.

Jenkins, R. (2004). Globalization, production, employment and poverty: Debates and evidence. *Journal of International Development, 16*(1), 1—12. https://doi.org/10.1002/jid.1059.

Johnson, C. (2007a). Impacts of prefabricated temporary housing after disasters: 1999 earthquakes in Turkey. *Habitat International, 31*(1), 36—52. https://doi.org/10.1016/j.habitatint.2006.03.002.

Johnson, C. (2007b). Strategic planning for post-disaster temporary housing. *Disasters, 31*(4), 435—458. https://doi.org/10.1111/j.1467-7717.2007.01018.x.

Johnson, C., Lizarralde, G., & Davidson, C. H. (2006). A systems view of temporary housing projects in post-disaster reconstruction. *Construction Management & Economics, 24*(4), 367—378. https://doi.org/10.1080/01446190600567977.

Kalin, M. E. (2005). The death toll from natural disasters: The role of income, geography, and institutions. *The Review of Economics and Statistics, 87*(2), 271–284. https://doi.org/10.1162/0034653053970339.

Kar, N. B. (2012). Impact of globalization on urban slum dwellers-migrants from rural areas: A social case study of Barasat Town, North 24-Parganas District, West Bengal. *Society Today: An Interdisciplinary Journal of Social Sciences, 1*(1).

Kelly, P. M., & Adger, W. N. (2000). Theory and practice in assessing vulnerability to climate change and facilitating adaptation. *Climatic Change, 47*(4), 325–352. https://doi.org/10.1023/A:1005627828199.

Kleinen, J. (2007). Historical perspectives on typhoons and tropical storms in the natural and socio-economic system of Nam Dinh (Vietnam). *Journal of Asian Earth Sciences, 29*(4), 523–531. https://doi.org/10.1016/j.jseaes.2006.05.012.

Leach, M., Mearns, R., & Scoones, I. (1997). Environmental entitlements: A framework for understanding the institutional dynamics of environmental change. Brighton: Institute of Development Studies, University of Sussex.

Lizarralde, G. (2015). The invisible houses: Rethinking and designing low-cost housing in developing countries. New York, NY: Routledge.

Lizarralde, G., Johnson, C., & Davidson, C. (Eds.). (2010). *Rebuilding after disasters: From emergency to sustainability.* New York, NY: Spon Press.

Miller, F., Osbahr, H., Boyd, E., Thomalla, F., Bharwani, S., Ziervogel, G., … Rockström, J. (2010). Resilience and vulnerability: Complementary or conflicting concepts? *Ecology and Society, 15*(3), 11.

Mutisya, E., & Yarime, M. (2011). Understanding the grassroots dynamics of slums in Nairobi: The dilemma of Kibera informal settlements. *International Transaction Journal of Engineering, Management, & Applied Sciences & Technologies, 2*(2), 197–213.

Norris, F. H., Stevens, S. P., Pfefferbaum, B., Wyche, K. F., & Pfefferbaum, R. L. (2008). Community resilience as a metaphor, theory, set of capacities, and strategy for disaster readiness. *American Journal of Community Psychology, 41*(1), 127–150. https://doi.org/10.1007/S10464-007-9156-6.

Pelling, M. (2003). *The vulnerability of cities: Natural disasters and social resilience.* London: Earthscan.

Sassen, S. (2004). The Global City: Introducing a concept. *The Brown Journal of World Affairs, 11*(2), 27–43.

Sen, A. K. (1982). Poverty and famines: An essay on entitlement and deprivation. Oxford: Oxford University Press.

Sen, A. K. (1997). *Resources, values, and development.* Boston, MA: Harvard University Press.

Taylor, P. J., Walker, D. R., Catalano, G., & Hoyler, M. (2002). Diversity and power in the world city network. *Cities, 19*(4), 231–241. https://doi.org/10.1016/S0264-2751(02)00020-3.

UNISDR. (2005). *Hyogo Framework for Action 2005–2015: Building the resilience of nations and communities to disasters.* Retrieved from https://www.unisdr.org/files/1037_hyogoframeworkforactionenglish.pdf.

UN-Habitat. (2003). *The challenge of slums: Global report on human settlements 2003.* London: Earthscan.

UN-Habitat. (2007). *UN-habitat for safer cities 1996-2007.* Nairobi, Kenya: UN-Habitat.

UNISDR. (2013). *UNISDR terminology on disaster risk reduction.* Retrieved from http://www.unisdr.org/we/inform/terminology.

Watts, M. (1983). On the poverty of theory: Natural hazards research in context. In K. Hewitt (Ed.), *Interpretation of calamity: From the viewpoint of human ecology* (pp. 231–262). Boston, MA: Allen & Unwinn.

Index

'Note: Page numbers followed by "f" indicate figures and "t" indicates tables.'